THE GOLDEN TICKET

THE GOLDEN TICKET

P, NP,
AND THE SEARCH
FOR THE IMPOSSIBLE

LANCE FORTNOW

PRINCETON UNIVERSITY PRESS

PRINCETON AND OXFORD

press.princeton.edu

Fourth printing, first paperback printing, 2017

Paperback ISBN: 978-0-691-17578-2

The Library of Congress has cataloged the cloth edition as follows:

Fortnow, Lance, 1963–
The golden ticket : P, NP, and the search for the impossible / Lance Fortnow.
pages cm

Summary: "The P-NP problem is the most important open problem in computer science, if not
all of mathematics. The Golden Ticket provides a nontechnical introduction to P-NP, its rich
history, and its algorithmic implications for everything we do with computers and beyond. In
this informative and entertaining book, Lance Fortnow traces how the problem arose during
the Cold War on both sides of the Iron Curtain, and gives examples of the problem from a
variety of disciplines, including economics, physics, and biology. He explores problems that
capture the full difficulty of the P-NP dilemma, from discovering the shortest route through all
the rides at Disney World to finding large groups of friends on Facebook. But difficulty also has
its advantages. Hard problems allow us to safely conduct electronic commerce and maintain
privacy in our online lives. The Golden Ticket explores what we truly can and cannot achieve
computationally, describing the benefits and unexpected challenges of the P-NP problem"—
Provided by publisher.

Includes bibliographical references and index.
ISBN 978-0-691-15649-1 (hardback)
1. NP-complete problems. 2. Computer algorithms. I. Title.
QA267.7.F67 2013
511.3'52—dc23
2012039523

British Library Cataloging-in-Publication Data is available

This book has been composed in Minion Pro

Printed on acid-free paper. ∞

Printed in the United States of America

5 7 9 10 8 6 4

To Marcy, Annie, and Molly,

SO THAT THEY MAY KNOW WHAT I DO AND WHY I DO IT.

In England the famous scientist Professor Foulbody invented a machine which would tell you at once, without opening the wrapper of a candy bar, whether or not there was a Golden Ticket hidden underneath. The machine had a mechanical arm that shot out with tremendous force and grabbed hold of anything that had the slightest bit of gold inside it and, for a moment, it looked like the answer to everything.

—Roald Dahl, *Charlie and the Chocolate Factory*

- - - - *Contents* - - - -

- - - - *Preface* - - - -

NEARLY HALF OF AMERICANS CARRY A SMARTPHONE, a computer more powerful than the supercomputers of just a couple of decades ago. Computers bring us the world's information and help us sort through it. Computers allow us to communicate with almost anyone, anywhere. Computers can perform tremendous computations, from simulating cosmic events to scheduling complex airline routes. Computers can recognize our voices, our faces, our movements. Computers can learn our preferences and tell us what books, music, and movies we may like. In the not too distant future, computer technology will allow our cars to drive themselves. There seems to be no limit to what computers can do.

Or is there? This book explores the computational problems that we may never be able to compute easily. It does so through the most important challenge in computer science, if not all of mathematics and science, the oddly named P versus NP problem.

P versus NP is a mathematical challenge, one of seven recognized as a Millennium Prize Problem by the Clay Mathematics Institute, which puts a million-dollar bounty on its solution. But P versus NP means so much more.

P refers to the problems we can solve quickly using computers. NP refers to the problems to which we would like to find the best solution. If P = NP, then we can easily find the solution to every problem we would like to solve. If P = NP, then society as we know it would change dramatically, with immediate, amazing advances in medicine, science, and entertainment and the automation of nearly every human task.

If P ≠ NP, by contrast, then there are some problems we cannot hope to solve quickly. That's not the end of the story, as we can create techniques that help us attack these problems in many cases. P ≠ NP means there is no automated way to solve some of the problems we want to solve. Still, knowing which tools don't work can help us figure out which ones do.

In August 2008, Moshe Vardi, editor-in-chief of the *Communications of the ACM*, asked me to write an article on the P versus NP problem. The Association for Computing Machinery is a major society serving computing researchers and professionals, and *Communications* is the society's main magazine devoted to articles of interest for that community.

At first I tried to push the article onto another computer scientist, but eventually relented. As Moshe put it to me, "If physicists write popular articles (and books) about string theory, we should be able to explain what complexity theory has accomplished, I'd hope." I wrote the article, aiming for the *Communications* audience, not just about the status of the P versus NP problem, which can be summarized as "still open," but about how people deal with hard problems. "The Status of the P versus NP Problem" was published in the September 2009 issue and quickly became the most downloaded article in the *Communications'* history.

The P versus NP problem remained a story to be told, and the popularity of the article suggested the time was right to tell this story, not just to scientists but to a much broader audience.

I used that short article as a framework for this book. Sections of the article become chapters here. I also took inspiration from Stephen Hawking's *A Brief History of Time*, which explains physics not through formulas and technicalities but with examples and stories. I attempt to do the same here to explore the spirit and importance of the P versus NP problem.

You will not find a formal definition of the P versus NP problem here; there are many great textbooks and websites that explore the definition of and technical results related to P versus NP. Reading this book will instead give you an appreciation of the possibilities and limits of computations as computers become such an integral part of our world.

Onward to P and NP!

Lance Fortnow
Evanston, Illinois

THE GOLDEN TICKET

Chapter 1

— — — — — — — — —

THE GOLDEN TICKET

A CANDY MANUFACTURER DECIDES TO RUN A CONTEST and places a handful of golden tickets inside its chocolate bars, hidden among the tens of millions of bars produced each year. The finders of these tickets will get a rare factory tour.

How do you find those tickets? You could buy up as many of the chocolate bars as you can. You can try using a magnet, but gold is not magnetic. Or you could hire thousands of people and give each of them a small stack of chocolate to sift through. It sounds silly, but not for Veruca Salt, who really wanted a golden ticket to visit Willie Wonka's chocolate factory.

Veruca's father, a rich business man, decided to buy up all the chocolate bars that he could find. This wasn't enough—just because you have a large mound of chocolate bars doesn't make the ticket any easier to find. Mr. Salt also had a factory and that factory had workers, and he wasn't afraid to use them to find that golden ticket hidden in the chocolate bars. He recounted how he found the ticket to the press:

> I'm in the peanut business, you see, and I've got about a hundred women working for me over at my joint, shelling peanuts for roasting and salting. That's what they do all day long, those women they sit there shelling peanuts. So I says to them, 'Okay girls', I says, 'from now on, you can stop shelling peanuts and start shelling the wrappers off these crazy candy bars instead!' And they did. I had every worker in the place yanking the paper off those bars of chocolate full speed ahead from morning till night.

"But three days went by, and we had no luck. Oh, it was terrible! My little Veruca got more and more upset each day, and every time I went home she would scream at me, '*Where's my Golden Ticket! I want my Golden Ticket!*' And she would lie for hours on the floor, kicking and yelling in the most disturbing way. Well, sir, I just hated to see my little girl feeling unhappy like that, so I vowed I would keep up the search until I'd got her what she wanted. Then suddenly . . . on the evening of the fourth day, one of my women workers yelled, 'I've got it! A Golden Ticket!' And I said, 'Give it to me, quick!' and she did, and I rushed it home and gave it to my darling Veruca, and now she's all smiles, and we have a happy home once again."

Like Mr. Salt, no matter how you try to find that ticket, you will need a considerable amount of time, money, or luck. Maybe someday someone will develop a cheap device that will let you find that ticket quickly, or maybe such a device will never exist.

Now, ten million is a very small number to today's computers. If you digitize the candy bars into a database, a typical desktop computer could search this database in a fraction of a second. Computers work much faster than a human searching through candy, but the problems are also much bigger.

What's the largest digital collection of data we have? How about the entire Internet? When you include all the video, audio, emails, and everything else, you have about 1,000,000,000,000,000,000 bytes of information, give or take a couple of zeros. A byte is about one character typed from your keyboard. That number sounds very large, but remember, computers are also very fast. A typical laptop can process about a trillion operations a second, which means you could theoretically search the entire Internet in just under four months if you could store the entire Internet in your laptop's memory. Google, with its hundreds of thousands of fast computers, scours the Internet constantly.

If computers can search the entire Internet quickly, then it sounds like we've solved the digital version of finding the golden ticket. But often we need computers not just to search the data we already have but to search for possible solutions to problems.

Consider the plight of Mary, a traveling salesman working for the US Gavel Corporation in Washington, D.C. Starting in her home city, she needs to travel to the capitals of all the lower forty-eight states to get the

Total Distance = 11,126

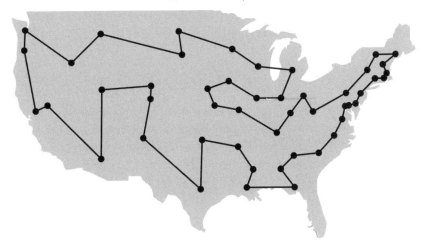

Figure 1-1. Traveling Salesman Problem Map.

state legislatures to buy an original US Gavel. US Gavel needs to reduce its travel expenses and asked Mary to find the best route through all the capitals that requires the smallest distance possible. Mary sketched a simple drawing on a map, played with it for a while, and came up with a pretty good route.

The travel department wanted Mary to see if she could come up with a different route, one that used less than 11,000 total miles. Mary wrote a computer program to try all possible different permutations of cities to find the quickest route, but a week later the program still hadn't finished. Mary sat down and did some calculations. There are forty-eight possible cities she could go to first, then she could choose among forty-seven of the remaining cities, then among the group of forty-six yet unvisited, and so on. That means a total of 48 × 47 × 46 × ... × 2 × 1 possible routes, equal to the sixty-two-digit number

12,413,915,592,536,072,670,862,289,047,373,375,038,521,486,354,677, 760,000,000,000.

If a computer could process one route in the time it takes light to cross the width of the smallest atom (about 0.0000000000000000033 seconds), it would still take about ten trillion trillion times the current age of the universe to check all of them. No wonder Mary's laptop hadn't finished in a week. Mary wondered whether there was some better way to find the best route, to find that golden ticket among the candy bars of possible trips.

That's the basic question of this book. The P versus NP problem asks, among other things, whether we can quickly find the shortest route for a traveling salesman. P and NP are named after their technical definitions, but it's best not to think of them as mathematical objects but as concepts. "NP" is the collection of problems that have a solution that we want to find. "P" consists of the problems to which we can find a solution quickly. "P = NP" means we can always quickly compute these solutions, like finding the shortest route for a traveling salesman. "P ≠ NP" means we can't.

The Partition Puzzle

Consider the following thirty-eight numbers:

14,175, 15,055, 16,616, 17,495, 18,072, 19,390, 19,731, 22,161, 23,320, 23,717, 26,343, 28,725, 29,127, 32,257, 40,020, 41,867, 43,155, 46,298, 56,734, 57,176, 58,306, 61,848, 65,825, 66,042, 68,634, 69,189, 72,936, 74,287, 74,537, 81,942, 82,027, 82,623, 82,802, 82,988, 90,467, 97,042, 97,507, 99,564.

These thirty-eight numbers sum to 2,000,000. Can you break them into two groups of nineteen numbers, where each group of numbers sums to 1,000,000? Feel free to use your calculator or spreadsheet, or write a computer program. (The solution is given at the end of the chapter.)

Not so easy, is it. There are over 17 billion different ways to break these numbers into two groups. With some careful coding on today's fast computers you can find a solution. But suppose I gave you 3,800 different numbers or 38 million. No way will a simple computer program give you the answer now!

Is this just a silly math puzzle? What if there was some clever computer procedure that would let us quickly figure out how to break a list of numbers into two groups with the same sum if those groups exist? If so, we could use this procedure to do much more than solve these math puzzles. We could use it to solve everything, even to find a traveling salesman's shortest route. This simple puzzle captures the P versus NP problem: any program that can solve large versions of this puzzle can compute just about anything.

The Hand

Your hands are the most incredible engineering devices ever created. Your hands can poke, grab, and point. Your hands can tie shoes or shoot an arrow. Hands can play a piano or violin, perform tricks of magic, precisely steer a car, boat, train, or plane. Your hands can shake someone else's hands or thumb wrestle with them. Hands can speak through sign language or by the words they write or type. Hands can softly caress or violently attack. They can use the delicate tools of a watchmaker or manipulate a chain saw. The right hands can create great works of art, or music, or poetry. Nearly everything humans have achieved, they've achieved through their hands.

The hand has twenty-seven bones; five fingers, including the all-important thumb; and a complex arrangement of nerves, tendons, and muscles, all covered with a flexible skin. But this incredible device, a marvel of nature's engineering, can do nothing on its own. The hand performs only according to instructions it gets from the brain. A dead man's hands tell no tales, or do much of anything else.

The hand is nature's hardware, and hardware on its own cannot do much. The hand needs software, messages from the brain that tell the hands how to perform and accomplish the task the brain wishes the hands to do.

Yoky Matsuoka, a robotics professor at the University of Washington, runs a group that developed an anatomically correct robotic hand. The fingers have the full range of motions and movements that human fingers do. Physically, her robotic hand can perform any of the incredible things our hands can do, but in reality the robotic hand doesn't do more than very simple tasks. Writing computer programs to control all the various aspects of Matsuoka's hand is a complex task. Coordinating different muscles means even the simplest tasks require complex code.

Somehow our brain manages to control our hands. The brain is mostly a high-powered computer. If our brains can do it, there must also be computer programs that tell these hands how to tie shoes and execute great art.

Knowing that such programs exist and finding them are two different issues. In time, computer scientists will develop more and more sophisticated programs, and Matsuoka's hands will perform more complex activities. Perhaps someday those hands may perform tasks beyond what human hands can do. It will be an exciting journey, but likely a very slow one.

Does it have to be? Suppose we could simply describe a task and immediately have a program that provided that functionality. Feed in a movie of a human tying a knot, and immediately have the computer repeat the process with robotic hands. Give a computer the complete works of Shakespeare, and have that computer create a new "Shakespeare" play. Suppose whatever we can recognize we can find. We can if P = NP.

That's the excitement of the P versus NP problem. Can everything be made easy, or do we need to do some things the hard way? We can't rule it out. Nevertheless, we don't expect life to be so easy. We don't think that P = NP, but the possibility of such a beautiful world is tantalizing.

P versus NP

P versus NP is about all the problems described above and thousands more of a similar flavor: How fast can we search through a huge number of possibilities? How easily can we find that "golden ticket," that one best answer?

The P versus NP problem was first mentioned in a 1956 letter from Kurt Gödel to John von Neumann, two of the greatest mathematical minds of the twentieth century. That letter was unfortunately lost until the 1980s. The P versus NP problem was first publicly announced in the early 1970s by Steve Cook and Leonid Levin, working separately in countries on opposite sides of the Cold War. Richard Karp followed up with a list of twenty-one important problems that capture P versus NP, including the traveling salesman problem and the partition puzzle mentioned earlier. After Karp, computer scientists started to realize the incredible importance of the P versus NP problem, and it dramatically changed the direction of computer science research. Today, P versus NP has become a critical question not just in computer science but in many other fields, including biology, medicine, economics, and physics.

The P versus NP problem has achieved the status of one of the great open problems in all of mathematics. Following the excitement of Andrew Wiles's 1994 proof of Fermat's Last Theorem, the Clay Mathematics Institute decided to run a contest for solutions to the most important unsolved mathematical problems. In 2000, the Clay Institute listed seven Millennium Problems and offered a $1 million bounty for each of them.

1. Birch and Swinnerton-Dyer conjecture
2. Hodge conjecture
3. Navier-Stokes equations
4. P versus NP
5. Poincaré conjecture
6. Riemann hypothesis
7. Yang-Mills theory

One of the Millennium Problems, the Poincaré conjecture, was solved by Grigori Perelman in 2003, though he has declined the $1 million prize. The other six, as of this writing, are unsolved.

Solve the P versus NP problem and get $1 million, a truly golden ticket!

Even better if you can show P = NP, for you then have a procedure that finds golden tickets, such as the solutions to all the other Millennium Prize Problems. Prove P = NP and get $6 million for solving the six unsolved Millennium Problems. Showing either P = NP or P ≠ NP won't be easy. If you want $6 million, you'll have a better chance playing the lottery.

Finding the Ticket

Sometimes we can find that golden ticket. Suppose I get in my car in Chicago and want to head to New York City. I plug the address into my GPS, which in a minute or two will find the fastest route from Chicago to New York, and off I go. The full street map of the United States can fit in a few million bytes of memory, but the number of possible routes described by this map is much higher. How many routes are there between Chicago and New York? Even if we require the car to not go in the wrong direction, a conservative back-of-the-envelope calculation gives more than a vigintillion, or 1 followed by sixty-three zeros, different possible routes. Yet the GPS still finds the quickest route even though it doesn't have time to look through all these possibilities.

Travel times have an interesting property. Pick an intermediate location, say, Pittsburgh. The shortest trip from Chicago to New York through Pittsburgh is just the shortest trip from Chicago to Pittsburgh followed by the shortest trip from Pittsburgh to New York. There are faster routes that avoid Pittsburgh, but the shortest trip from Chicago to New York can't do worse than the shortest trip from Chicago to Pittsburgh and Pittsburgh to New York.

The GPS uses computer programs based on these properties to quickly narrow down the best route. The GPS may still need to examine tens or hundreds of millions of routes but nothing a computer processor can't handle, as opposed to the vigintillion number of total possibilities.

Finding the shortest path doesn't capture the full power of P versus NP. The shortest path problem tells us that just because there are a huge number of possibilities, we don't always need to explore them all. P versus NP asks us if we ever need to explore all the possibilities of any search problem.

The Long Road

In this book we tell the story of P and NP. What are P and NP? What kinds of problems do they capture? What are NP-complete problems, the hardest of all search problems? How do these problems capture the P versus NP question?

One simple example: What is the largest clique on Facebook, that is, the largest group of people all of whom are friends with each other? Is it a hundred people? A thousand? Even if you had access to all of

Facebook's data, this could be a difficult problem to solve. Finding large cliques is as hard as every search problem.

What happens if P = NP? We get a beautiful world where everything is easy to compute. We can quickly learn just about everything, and the great mysteries of the world fall quickly, from cures to deadly diseases to the nature of the universe. The beautiful world also has a dark underbelly, including the loss of privacy and jobs, as there is very little computers cannot figure out or accomplish.

The beautiful world is an unlikely outcome. That leaves us with hard search problems we still want or need to solve. We don't always need to give up. Computer scientists have developed many techniques, from heuristics that solve many of the problems most of the time to approximation techniques that give us close to ideal solutions.

How did we get to P and NP? This great story is actually two separate stories, taking place at a time when the world was divided by the Cold War. The ideas and questions of efficient computation were developed separately in those worlds, with two tales that lead to the same place, the P versus NP question.

How do we approach a proof that P ≠ NP? Kurt Gödel showed that mathematics cannot solve every problem. Can similar techniques show that there are search problems we can't solve quickly? Perhaps we can break computation down into its simplest components to analyze the difficulty of solving problems. Algebraic geometry, an abstract branch of mathematics, gives us some new hope, but we remain very far from settling this important question.

What good can come out of P ≠ NP? It can help us keep secrets and make fake random numbers that look really random.

Can future computers based on quantum mechanics make the P versus NP problem irrelevant? Not likely, but if built, they could solve some problems, like factoring large numbers, that remain out of reach for current machines. Quantum mechanics can also give us unbreakable secrets whether or not P = NP.

What about the future? The great challenges of computing still lie ahead of us. How do we deal with computers that have to work together to solve problems? How do we analyze the massive amount of data we generate every day? What will the world look like when we network everything? The P versus NP problem only grows in importance as we try to tackle these challenges ahead.

Partition Puzzle Solution

The thirty-eight numbers

14,175, 15,055, 16,616, 17,495, 18,072, 19,390, 19,731, 22,161, 23,320, 23,717, 26,343, 28,725, 29,127, 32,257, 40,020, 41,867, 43,155, 46,298, 56,734, 57,176, 58,306, 61,848, 65,825, 66,042, 68,634, 69,189, 72936, 74,287, 74,537, 81,942, 82,027, 82,623, 82,802, 82,988, 90,467, 97,042, 97,507, 99,564

can be split into the following two groups:

15,055, 16,616, 19,390, 22,161, 26,343, 40,020, 41,867, 43,155, 46,298, 57,176, 58,306, 65,825, 66,042, 69,189, 74,537, 81,942, 82,623, 82,988, 90,467

and

14,175, 17,495, 18,072, 19,731, 23,320, 23,717, 28,725, 29,127, 32,257, 56,734, 61,848, 68,634 72,936, 74,287, 82,027, 82,802, 97,042, 97,507, 99,564.

Each of these groups of numbers sums to 1,000,000.

Chapter 2

— — — — — — — — — —

THE BEAUTIFUL WORLD

IMAGINE YOU ARE ASKED TO WRITE AN ESSAY that captures the social changes caused by the Internet over the past twenty years. Will you write about a device in your pocket that gives you access to all public information instantaneously? Or about how social networks connect us in entirely new ways? Or will you talk about the massive changes in the music, movie, publishing, and news businesses? One essay cannot begin to do justice to the changes that have occurred over the past two decades. Now imagine writing that chapter in the 1990s before the changes actually happened.

If it turns out that P = NP and we have efficient algorithms for all NP problems, the world will change in ways that will make the Internet seem like a footnote in history. Not only would it be impossible to describe all these changes but the biggest implications of the new technologies would be impossible to predict.

To give you a tiny taste of this beautiful world, let's imagine a future world a few years after the discovery of an efficient algorithm that solves NP problems. Let's jump to the year 2026 and explore this beautiful world, almost surely a fantasy world, the world of P = NP. First let's see how this world developed.

The Urbana Algorithm

In 2016, a Czech mathematician, Milena Pavlova, emailed around a theoretical algorithm for solving NP problems efficiently. After careful verification by the computer science and mathematics communities, the consensus was that Milena's algorithm worked as advertised and the great P versus NP problem was solved. Milena published her algorithm in a paper under the modest title "On an Open Problem of Cook," but an article in the *New York Times* aptly described the result with the simple title "P = NP."

In 2018, Milena Pavlova was the first woman to receive a Fields Medal, the highest honor in mathematics. A year later the Clay Mathematics Institute awarded a check for $1 million to Milena. She became the second person to successfully solve one of the Clay Mathematics Institute Millennium Problems, after Grigori Perelman. Unlike Perelman, Milena happily took the prize money, though she did donate an undisclosed amount to fund scholarships at her home university in Prague.

While Milena's algorithm was a theoretical breakthrough, the algorithm itself still took too much time to have any practical value. But in 2017 a Russian computer scientist, Minsk Borov, discovered a clever twist to Milena's algorithm that improved the algorithm dramatically, although it still did not seem very practical.

A year later a group of undergraduate students at Tsinghua University optimized Minsk's code carefully and ran it on the world's fastest supercomputer, then in China. They were able to solve moderately large clique problems and several other common NP problems in a matter of days. Some major industrial companies, such as Boeing and Daimler-Benz, contracted with Tsinghua University to help solve some of their tricky optimization problems. Boeing got a better wing design for its new 797 aircraft, allowing it to fly from London to Sydney nonstop.

Steve Frank was a PhD student at the University of Illinois who was visiting Tsinghua for a semester during this time and was one of the researchers working on this project. After returning to Illinois in Urbana, he lamented to his adviser that despite all their optimization, it still took several days to solve even a single moderately sized NP problem.

"So what do you ask a genie who will grant you only one wish?" said the adviser.

"I have no idea," replied Steve.

"You ask for a genie who will grant all your wishes."

The proverbial light bulb went off in Steve's head. He knew there must be some better algorithm for solving clique problems out there somewhere, but he couldn't figure it out on his own. But he had the genie, the Tsinghua code, which could search an exponential number of possibilities quickly. So he wrote up a program that used the Tsinghua routines to search for a better algorithm for NP problems. He got permission to use the computing resources of the National Center for Supercomputing Applications (NCSA), based at the University of Illinois. After weeks of processing time his work paid off a little bit, finding a new algorithm that had a 5 percent improvement over the Tsinghua code—good enough for a research paper but not enough to make a real impact.

His advisr simply said, "Try again using the new code."

So Steve used his new code to find an even faster algorithm for NP problems. A few weeks later he had a 20 percent improvement.

But his adviser was not impressed. "Try it again."

Steve replied, "Why don't I just set up the computer to automatically keep trying with the new code it finds?"

The adviser gave that look, the look that told a student he had achieved enlightenment, or at least had realized the obvious.

Steve went back to his office and started the tricky process of writing code that searches for faster code, and then used this faster code to find even faster code and continue this process until it could find no further improvement. One of his officemates asked Steve if he was worried about the Skynet effect.

"Skynet?"

"You know, when a computer becomes so smart it becomes self-aware and takes over the world, like Skynet in the *Terminator* movies."

"No, this is just computer code. Don't worry."

For the final time, Steve ran his code on the NSCA's supercomputer. The computer generated better and better algorithms for NP problems. In the end it stopped with a program that had 42 million lines of unintelligible machine code. But it solved NP problems fast, very fast. (And no, the computer never became self-aware.) A university press release touted the excitement of this new Urbana algorithm, and the name stuck.

Some mathematicians at the University of Illinois with early access to the Urbana algorithm started using the algorithm to find (mostly incomprehensible) proofs of some of the other Millennium Problems. The Clay Mathematics Institute quickly issued a statement that it would not accept any new proofs based on any algorithm derived from the $1 million algorithm that Milena Pavlova had already developed.

Many companies wanted to license or outright buy the rights to the Urbana algorithm but ran into a legal quagmire—everyone from the Czech government that funded Milena's research, to Steve Frank, to Steve's adviser, to the NSCA claimed ownership of the Urbana algorithm. Realizing the algorithm's great importance, the World Trade Organization declared that the Urbana algorithm would be released into the public domain after proper compensation, which would be worked out by an arbitration committee. Only the Chinese objected, but they finally realized they were powerless to block the release, and on October 23, 2019, the Urbana algorithm became available to all.

And then the world changed. . . . .

Computers 1, Cancer 0

Helen's doctor walks into the examination room, closing the door behind him. He says, "I have some difficult news, you have pancreatic cancer".

Helen gasps. She is only forty-two and has a family with three kids, ages six to fifteen. "How do you know? I just had a blood test."

"We can determine whether or not you have cancer, what kind of cancer, and the extent of the cancer based solely on markers in your blood, combined with your DNA. We can do a biopsy if you insist, but the accuracy of these tests makes the biopsy procedure not worth the risk."

"My cousin was diagnosed with pancreatic cancer eight years ago, in 2018. There weren't many treatment options, and she passed away seven months later."

"Much has changed in the last decade. General approaches to cancer have had limited success. We realized that one needs an individual approach. We can examine a person's DNA as well as the mutated DNA of the cancer cells and develop proteins that will fold in just the right way to effectively starve the cancer cells without causing any problems for the normal cells. The dead cancer cells just wash out of the body."

"Sounds expensive," notes Helen.

"Chemotherapy was expensive. This will only cost a couple of thousand dollars, and even that will get much cheaper in the near future. Your insurance will cover most of it."

"Amazing! What happened in the past decade that made testing and treatment so simple?"

"We have had ideas along these lines for years, but even the world's fastest computers could only make limited breakthroughs on the DNA code. The development of the Urbana algorithm changed all that, and within the past few years we have made incredible progress. Normally we would have done years of experiments, but the initial trials were so successful the FDA felt it unconscionable to not allow immediate use."

"When do we get started?" asks Helen.

"Started? We already did the analysis from the blood that we drew. Here is your prescription."

The doctor hands Helen not a piece of paper but a USB drive. "This drive has the encoding of the proteins that you need. Bring it to your pharmacy and they'll produce pills for you. One a day for the next two weeks should clear the cancer out of your system. And there shouldn't be any side effects. But remember, this prescription will work only for you. If anyone else takes these pills with their different DNA, the pills could have serious and possibly fatal consequences."

"No side effects? No hair falling out? No feeling nauseated? Early detection and I just take a pill and it goes away, just like a cold? All because of an algorithm?"

"Not quite," the doctor says. "Sure, the Urbana algorithm has helped us cure cancer, AIDS, and diabetes, but the common cold still remains a mystery."

The Baseball Game

"A perfect day for baseball!" Randy tells Kate, his twelve-year-old daughter, taking her to her first ball game in St. Louis. The Cardinals are hosting the Milwaukee Brewers in a tight pennant race game late in the 2026 season. Randy is thinking about how the game has changed since he was a kid, especially in recent years since the Urbana algorithm has had such a great effect on a game so technologically simple it doesn't even use a clock.

First, of course, is the schedule of this particular game. As late as 2004, a husband-and-wife team, Henry and Holly Stephenson, scheduled games for Major League Baseball. They used a few simple rules, like the number of games played at home and away by each team, and some local quirks, like the Boston Red Sox like to host a daytime baseball game on Patriot's Day in mid-April, as the runners in the Boston Marathon pass nearby. In 2005, Major League Baseball contracted with a Pittsburgh company, the Sports Scheduling Group, because its scheduling could better avoid teams playing each other in consecutive weeks. Although you can't play baseball outdoors during a rainstorm, the Sports Scheduling Group never considered rain when making the schedule for the simple reason that one could not predict rain when the schedule was announced in the December preceding the season. Of course, that was before the Urbana algorithm.

The algorithm has made incredible advances in weather prediction, allowing accurate predictions of temperature, winds, cloud cover, and precipitation nearly a year ahead of time. Similar algorithms now save lives by accurately predicting storms, tornados, and hurricanes so people can prepare or evacuate as needed. But new weather predictions have also altered everyday life as well. Schools plan snow days in advance in their schedules. Outdoor wedding chapels charge extra for great weather days and give discounts for those willing to get married on a hot, humid, or rainy day. More people come to baseball games on good weather days, so it makes no sense to have a game in rainy or even cloudy Detroit while the sun shines over an empty stadium in Minneapolis.

Both the fans in the stadium and the television audiences want to see games that really matter, the best teams playing each other late in the

season. Back when Randy was young, which teams would really succeed was more guesswork than science. Newer prediction tools can un-
cannily tell which teams will be in contention in September. They don't
always predict the winners; the randomness of baseball can add a few
wins or losses a season. These models do predict which teams will be
close to first place late in the season. Every year the fans of teams not
expected to compete say this will be the year the computer is wrong and,
with one famous exception, every year it isn't.

So the baseball czars want to schedule games in a way that everyone
has the best possible weather and good teams play each other at the end
of the season, not to mention more mundane cost savings like reducing the amount of travel for each team. Handling all these issues and
the multitude of possible schedules would have been impossible a mere
fifteen years ago, but the Urbana algorithm spits out the best schedule in
a matter of minutes.

So Randy gets to take Kate to an important game in perfect weather.
Of course, the Cardinals charge extra for this date, but Randy is quite
willing to pay for the privilege.

As Randy and Kate walk into the stands, there are no ticket takers
or turnstiles. They just walk into the stadium. A small camera watches,
and face recognition software matches ticket owners with their faces. If
someone without a ticket tries to enter, security will find that person,
but most people have long since stopped trying to beat the software.

Randy and his daughter take their seats. Randy looks around and
sees an empty platform where a large TV camera and its operator would
have stood just a few years ago. Instead, twenty small, nearly invisible
cameras are set up around the field. Computer software based on the
Urbana algorithm will take these twenty camera feeds and instantly create a rich 3-D rendering of the play of the game. From this rendering the
computer can construct a video that shows the play from any location
on the field in any direction. One can see a pitch from the eyes of the
catcher or the runner's point of view as he slides into third. Technically,
these images are computer-generated, but they look quite real to anyone watching. In a famous experiment one hundred people were shown
an actual video feed side by side with the computer-generated picture.
Eighty-nine of them thought the computer picture was the real one.

Some of the newer televisions can directly download the 3-D rendering, allowing the TV viewer to virtually fly anywhere in the ballpark as the game is going on.

As soon as the bat hits the ball, the computer predicts exactly where the ball will go, which allows the computer to pick the exact location to get an optimal view of the play as it occurs. The TV network needs just four people, two commentators, a producer, and a technician for when things go wrong, which rarely happens. The commentators speak in English. St. Louis and Milwaukee each have a well-known Japanese player, so this game is watched closely in Tokyo, where viewers hear Japanese commentary. Actually, the Japanese hear the American announcers, but the computer instantly uses voice recognition, translation, and voice generation tools based, naturally, on the Urbana algorithm to give Japanese commentary spoken like natives. The game is available in 876 languages and dialects across the world.

A vice president at the network wonders if they even need four people working the game. The right number might be zero. For years, college campuses have had cameras around their playing fields to record the games for the coaches to analyze later. An entrepreneurial student recently figured out how to use the Urbana algorithm on these camera feeds to create full-fledged broadcasts complete with commentary, well-chosen angles, statistics, and instant replays entirely generated by computer. It wasn't long before every college game (and many high school and even Little League games) in every sport was available to watch in real time over the Internet in nearly every language. The broadcasts don't quite have the quality of real people doing the commentary, but they aren't that bad.

Fantasy sports has reached a new level such that one can create computer-generated games with players from different teams or even players from different eras: Cy Young pitching to Joe DiMaggio, both in their prime, or the 1927 Yankees playing against the 1998 Yankees. It is difficult to separate the fantasy from the reality. One April Fools' Day, a computer programmer at ESPN rewrote some code so that everybody watching the TV broadcast saw his favorite basketball team, the Boston Celtics, beat the New York Knicks even though the Knicks won the game on the real court. Many heads rolled after that disaster.

The computer can also act as a perfect umpire, always correctly calling balls and strikes, outs, and home runs with uncanny accuracy. The minor leagues and colleges now use these computer umpires for all their games to cut costs, but the Major Leagues insist on real umpires, a point of controversy after every game-changing bad call.

Fearing, correctly, that the computer could now outmanage any human manager, Major League Baseball banned the use of computer devices by the teams during a game after the 2022 World Series debacle. "For the good of the game," the commissioner claimed.

As Randy drinks his beer and eats his hot dog (both better-tasting and healthier with new recipes developed by the Urbana algorithm), he watches the game on the field just as it was played when he went to the game as a kid. He could have saved considerable money and watched the game with perfect views and glorious 3-D at home. But even as technology has caused the home experience to far surpass what one gets at the stadium, Randy loves being part of it, an emotional attachment that even the Urbana algorithm has failed to recreate.

The computer can generate a box score, but Randy still teaches Kate how to keep score on paper, as his father did for him. Technology, particularly the Urbana algorithm, changes almost everything, but the game remains the same. It will always be three strikes and you're out at the old ball game.

Occam's Razor

How does simple computer code, even as powerful as the Urbana algorithm, give us a world where we can cure diseases, predict the weather, and create imaginary environments indistinguishable from true reality? To answer that question we need to go back to the early fourteenth century.

William of Ockham joined the English Franciscan order at a young age. The Franciscans and many other religious orders back in the fourteenth century settled in Oxford and maintained housing for students attending the University of Oxford. As one of these students, William studied theology, but never graduated. Nevertheless, he would go on to become one of the major thinkers of the Middle Ages, making great contributions to physics, theology, logic, and philosophy. We know him best for Occam's razor, the principle that the simplest explanation is usually the best, denoted a "razor" supposedly because it allows us to "shave off" the complicated parts of a theory, leaving the simple explanation behind. This philosophy guided scientific and philosophical thinking throughout the Renaissance and continues to guide us today.

René Descartes, a seventeenth-century French philosopher, used the Occam razor principle to argue even for the existence of the world around him. Descartes is, of course, most famous for his philosophical statement "Cogito ergo sum," or "I think, therefore I am." In his treatise *Discourse of the Method*, Descartes from assuming nothing deduces his own existence from the mere fact that he can reason about himself. What about the complex world that Descartes experiences? Could everything exist merely within Descartes' consciousness? Descartes rejects that idea for the simpler and more likely explanation that other people are just like him, living in this physical world we can study and attempt to understand.

> And as I observed that in the words "I think, therefore I am", there is nothing at all which gives me assurance of their truth beyond this, that I see very clearly that in order to think it is necessary to exist, I concluded that I might take, as a general rule, the principle, that all the things which we very clearly and distinctly conceive are true, only observing, however, that there is some difficulty in rightly determining the objects which we distinctly conceive.
>
> Descartes, *Discourse*, Part IV

A contemporary of Descartes, Johannes Kepler, examined planetary motion and derived a set of rules that described the path and speed of planets in their orbits. Kepler had no simple description of why planets followed these rules.

Isaac Newton applied the principle of Occam's razor to the physical world. Newton's famous laws of motion gave very simple rules on how objects react.

1. If no force is applied to an object, an object at rest remains at rest and one in motion remains at a constant velocity.

2. A force applied to an object creates a fixed acceleration in proportion to the size of the force.
3. Every force applied by one object to another has an equal and opposite force from the second object to the first.

Combined with his simple description of gravity, Newton showed how to derive Kepler's rules for planetary motion. The simple explanations can have great explanatory power.

Centuries later, Albert Einstein and others theorized that Newton's simple laws of motion break down when objects travel close to the speed of light, and experiments have mostly proved Einstein right. Einstein famously remarked, "Make everything as simple as possible, but not simpler." But that doesn't mean Newton was wrong. Rather, his models gave a good approximation to the world as he saw it and work well even today for simple activities such as driving cars or doing experiments in high school science labs.

Even Einstein's theories break down when dealing with very small particles, which seem to follow a set of rules known as quantum mechanics. Physicists today grapple with finding theories that will merge Einstein's general relativity with quantum mechanics, leading to a grand "theory of everything."

Simple models can never capture the entire complexity of the world, but they often get very close. Find a simple explanation of how things work and you can make good predictions of how similar things will work in the future. Recently we have seen this principle take hold in the computer science world, to great effect.

Today I can take a handwritten check, take a picture of it with my phone, and deposit the check over the Internet. The bank's software looks at the check and determines the amount as well as the account numbers. The system works even if the check is handwritten. No bank employee will ever look at the check if it is not disputed.

Recognizing the account numbers on the bottom of the check is pretty straightforward. These numbers follow a fixed format specifically designed to be easily read by computers.

But the amount on the check, $30.00, is handwritten. How can a computer determine the value of a check when everyone has different handwriting?

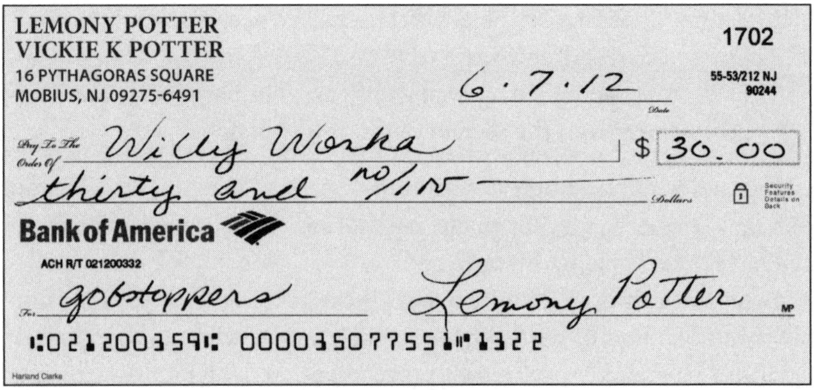

LEMONY POTTER
VICKIE K POTTER
16 PYTHAGORAS SQUARE
MOBIUS, NJ 09275-6491

1702

6 · 7 · 12

55-53/212 NJ
90244

Pay To The Order Of Willy Worka $ 30.00

thirty   and   no/100 ————————   *Dollars*

Bank of America

ACH R/T 021200332

For gobstoppers   Lemony Potter   MP

⑈021200359⑈ 00003507755⑈ 1702

Harland Clarke

Figure 2-1. Check.

Not an easy problem. Just look at the various ways people can write the number 2.

An area of computer science called machine learning tackles problems like these. First, the algorithms are trained on large amounts of data, in this case thousands of examples of the number 2 as well as all the other digits. The algorithm tries to fit a relatively simple model that properly classifies the handwritten digits. Properly trained these algorithms will classify nicely new numbers, even those written well after the training phase.

Figure 2-2. Twos.

In the past twenty years, computer scientists have made great strides in machine learning, developing techniques that can examine thousands to millions of examples and figure out the right way to classify data. Beyond check recognition, many photo programs will do a passable job of sorting pictures by people's faces. Companies like Amazon, Netflix, and Pandora can recommend books, movies, and music based on your previous purchases, as well as on your watching and listening patterns. Voice recognition and language translation programs won't fool a human but will give you a reasonable idea of what was written or said. Email spam detection has virtually eliminated unwanted messages, and cars are expected to all but drive themselves by 2020.

These techniques can only take us so far. Advances in the future may eventually become incremental. Does this mean Occam's razor has met its limits?

Not really. While the principle of Occam's razor holds that the simplest way to describe the data is the best one, it doesn't tell you how to find that simple description. Current machine learning techniques can only use simple kinds of descriptions, ones that combine various features in relatively straightforward ways. Finding the shortest description, the smallest, efficient computer program of any type that can classify the data efficiently, is a difficult problem, an NP problem.

With the Urbana algorithm we can solve all the NP problems quickly, finding the simplest program that classifies data becomes an easy programming exercise. All we need do is feed in lots of data and the algorithm does the rest. And that lets us learn just about everything.

We've seen how these ideas can help cure diseases and change America's game. Let's go back to the future and see how the Urbana algorithm changes art itself.

Automating Creativity

Via Occam's razor, we can use the Urbana algorithm to learn just about anything, including what makes good art, popular music, and words that stir the soul. Remember that $P = NP$ means that what we can test, we can find. So once you have an algorithmic process to recognize greatness, you can use the Urbana algorithm again to quickly find that greatness.

Two weeks before the Democratic primary for the 2022 U.S. Senate race in Colorado, Pete Johnson placed a distant third in the polls. That was before "the speech." In a small theater in Vail with no press in attendance, Johnson gave a ten-minute speech about Colorado and America that stirred the soul. He got a rousing standing ovation from the thirty-two people who attended the speech.

One of the attendees recorded the speech on her cell phone and posted it to the Internet. Millions of people watched the speech online. Thousands of people would later claim to have seen the speech live. Johnson won the primary election, and there was serious discussion about his being a candidate in the 2024 presidential election. Pete Johnson wanted more exposure, so he fired his campaign manager and hired someone with a national reputation.

The embittered former campaign manager held a press conference explaining the truth behind the speech. The campaign needed a quick boost or Pete Johnson's candidacy would soon end. The manager hired a computer programmer, who downloaded tens of thousands of well-received speeches through the decades. The programmer then used the Urbana algorithm to develop a new speech based on current events in Colorado and around the world. A well-rehearsed Pete Johnson gave that computer-generated speech in Vail. The campaign manager recruited an audience member to record the speech and post it online.

A great debate raged. Some people were outraged. Others didn't see much difference between speeches written by speechwriters and speeches written by computers. Peter Johnson lost his momentum and the general election, but by 2026 most candidates were using computers to help or write their speeches as a matter of daily practice. Of course, when everyone gives great speeches, then no one does.

Some classical musicians are using the Urbana algorithm to finish famous incomplete works such as Puccini's opera *Turandot*, Mahler's Tenth Symphony, and Schubert's Eighth Symphony, the *Unfinished*, as well as new works such as Beethoven's Tenth. The most famous symphony developed by the algorithm, dubbed Urbana's First, has a whole new sound loved by the concertgoing public. Some people even use the

Urbana algorithm to create new Beatles and Elvis records, even replicating the original voices. In all these cases real music critics claim that these pieces are all derivative and lack any real creativity, but many download the songs just the same.

Others use computers to generate new art, novels, plays, and poetry. One movie lover recently created a romantic comedy starring Humphrey Bogart and Julia Roberts in their prime, which nearly convinced the movie-watching public that those two had traveled through time to film the movie together. Want to see *The Wizard of Oz* directed by Tim Burton? No problem.

Amazon goes well beyond recommending books. For the cost of a movie ticket, Amazon can produce a novel specifically written for you, matching your interests and tastes. NBC currently airs a "live" action-adventure television series created entirely by computer, with no need for screen writers or actors. The latest Xbox has a video game that immerses you in a story and a world created as you play it. Instead of a having fixed set of possible story lines, this game can allow any actions of the players and follow those actions to their conclusion. More than a few people have trouble separating their game play from their real (and less exciting) lives.

The world loves all these amazing entertainment options. Many critics write about the loss of true expression. Whether the Urbana algorithm marks the birth of a new era in the arts or the death of creativity will remain a debate for centuries to come.

The Ultimate Detective

Law enforcement found the Urbana algorithm an incredible tool in solving crimes, seemingly doing the impossible in tracking down suspects. It did, however, cause some controversy.

In a multiple-murder case in Atlanta in the post–Urbana algorithm future, police managed to get DNA from a fast-food wrapper at the crime scene, but they could not find a match in the U.S. Combined DNA Index System (CODIS). A computer scientist at Georgia Tech used the Urbana algorithm on CODIS to create a new procedure that uses this DNA to predict not only physical features such as height, eye, and skin color but also personality traits and likely careers. The computer scientist and his students then matched CODIS data with mug shots and developed a technique to render a rough sketch of what a person looked like from his or her DNA, if you knew the person's age and type of facial hair.

The police then used the algorithm to match the DNA from the crime scene to mug shots but couldn't find a match. So they ran the algorithm on driver's license photos and had a suspect, George Brown, in custody within a few hours. Other than lacking an alibi and matching the computer-generated sketch, George Brown had no obvious connection to the murder. He refused to give his DNA, but the police got a court order, and the DNA matched that obtained at the crime scene, as expected. George Brown was found guilty and sentenced to death.

Brown appealed, and his lawyers argued that using the crime scene DNA to create a sketch violated George Brown's privacy rights. The case went to the U.S. Supreme Court, which ruled that the use of DNA to determine a person's identity is legitimate. The police did nothing more than run an algorithm on evidence they had legally acquired.

Still, the case caused a huge outcry from the public over loss of privacy. The decision caused thirty-eight states to ban the use of DNA other than to match identities already collected legitimately in CODIS. The governor of Georgia would eventually commute George Brown's death sentence to life in prison.

The Dark Side of the Beautiful World

The Urbana algorithm immediately broke public-key cryptography, the ability to send your credit card number securely to companies over the Internet without any initial setup. At first this caused some hiccups in electronic commerce, but the major Internet companies got together and soon moved to an older private-key system. A registry of private keys was quickly established, and you can now cheaply purchase a USB drive filled with billions of one-time use secret keys from your local drugstore. It is slightly more cumbersome than before, but electronic commerce hums along again.

The perceived loss of privacy scares a few more people. Not only can innocuous video cameras immediately recognize people and track them, algorithms can accurately predict the trips you take, the music you listen to, the movies you watch, and the products you purchase. The computers know more about you than you do. You receive advertising perfectly designed to change your shopping habits. Potential shoplifters and terrorists are identified almost immediately and closely watched.

The biggest concern about the Urbana algorithm is jobs. Almost every white-collar job from secretary to midlevel manager is affected by an algorithm that can take in information, even informally written emails, audio, and video, and produce letters and reports, draw conclusions, and make decisions. Jobs that were once outsourced are now Urbanaized, as it is called. Workers who cannot argue they add extra value find themselves out of work. Corporations continue to slash payrolls. Several companies are replacing their overseas help desks with computerized response systems that work in multiple languages and dialects. Many surveyed customers find them a more valuable resource than the humans they replace.

Some countries are setting up laws to hamper the use of the Urbana algorithms when they displace workers, but these laws rarely last long in the face of business competition from other countries. There are signs the situation is beginning to reverse itself. The Urbana algorithm is giving a boost to a sluggish economy, and whole new industries are starting to take shape. Universities are creating new courses in "Urbana optimization" that show how to take any problem and find the quickest and simplest way to use the Urbana algorithm to solve it. The algorithm will create far more jobs than were lost. Still, many feel anger against the algorithm on behalf of those people who lost their jobs and can't fit into the new economy.

Governments continue to pass laws to protect people from many of the consequences of the Urbana algorithm, but you can't put technology back in the bottle. Humanity has quickly adapted, and only a few outspoken people in polls would be willing to turn back the clock and go back to that ancient world of 2012, before an algorithm gave them the beautiful world.

Back to Reality

Sounds hard to believe that a single algorithm can change everything? That everything we can recognize we can find quickly? That we can learn everything easily? Yes, it does, which is why the vast majority of computer scientists believe that P ≠ NP, that there will never be an Urbana algorithm or anything like it.

Many of the stories in this chapter may yet come true no matter the fate of the P versus NP question. It might take a bit longer, and we'll have to tackle challenges individually as opposed to using magic computer code that solves all our problems. But human ingenuity is strong, and if we can dream it we can find a way to get there eventually.

Chapter 3

— — — — — — — — —

P AND NP

Frenemy

Nowhere can we understand P and NP better than in Frenemy, an imaginary world where every pair of people comprises either friends or enemies.

Frenemy has about 20,000 inhabitants. Every individual seems normal, but put two in close vicinity and a strange thing happens. Either the two take an instant liking to each other, immediately becoming the best of friends, or they take one look at each other and immediately become the worst of enemies. Despite the Frenemy name, two inhabitants are never observed taking a middle ground; they are always either close friends or distant enemies.

These relationships appear nearly random. Friends of your friends might be your friend or your enemy. Likewise, an enemy of an enemy might be a friend or not. There seems to be no connection with gender, race, religion, or social class, though people tend to have far fewer friends than enemies.

The Internet has produced tons of data about the friendship relationships in Frenemy. By examining social networking data such as Facebook and Twitter, computer scientists at the Frenemy Institute of Technology have put together a nearly complete database showing which pairs of people are friends and which pairs of people are enemies. In this chapter we will see what these researchers can and cannot do with this data.

Six Degrees

If we pick two arbitrary people in Frenemy, let's call them Alice and George, they are unlikely to be friends. But maybe there is an intermediary, Bob, who is friends with both Alice and George. Or maybe not. The scientists at the Frenemy Institute plotted all the people in Frenemy and put lines between those that were friends. Part of that diagram looked like this.

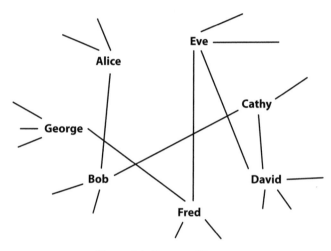

Figure 3-1. Frenemy Diagram.

To connect Alice and George, the researchers found six links to form a chain between them: Alice is friends with Bob, who is friends with Cathy, who is friends with David, who is friends with Eve, who is friends with Fred, who is friends with George. The scientists in Frenemy wondered whether every two people in Frenemy had a reasonably short chain of friendships connecting them, a "small world" phenomenon, named not after the Disney ride but because when two strangers meet and find some tenuous connection between them, they'll often exclaim, "It's a small world!"

In 1967 the psychologist Stanley Milgram ran a famous experiment to test the small world theory. He first picked a stockbroker in the Boston area. Milgram kept his name a secret, but let's call him Tom Jones. Milgram randomly chose about 100 stock holders from Nebraska, 100 other people in Nebraska, and 100 people solicited via a newspaper ad in the Boston area. The second group in Nebraska and the Boston group had no special connections to the investment world. For each of these people, Milgram provided a folder with instructions, a roster, and fifteen stamped reply cards addressed to Milgram at Harvard University. The instructions read:

1. Add your name to the roster.
2. Take one of the reply cards, fill it out, and stick it in the mail.
3. If you know the Boston stockbroker Tom Jones personally, then send this folder to him.
4. Otherwise think of someone not already on the roster that you know well and who is more likely than you to know the stockbroker, and send him or her the folder.

In all, 217 of the original 300 people sent out the document to their friends, and 64 of these letters eventually reached the stockbroker. The average chain length was 5.2 people, which led to the notion of six degrees of separation, that everyone is connected to everyone else through a chain length of just six people. Although many have criticized various aspects of Milgram's study, and Milgram himself would have never claimed the six-degrees rule, it certainly appears that we are all more closely connected than one would have expected.

If you use a better-defined notion of connectedness than just knowing someone, you can start to analyze real networks, which leads to some parlor games seeing how far one is from a specified highly connected person. In 1994 Kevin Bacon, while promoting his movie *The River Wild*, jokingly remarked that he had worked with everyone in Hollywood or had worked with someone who had worked with them. A trivia game, Six Degrees of Kevin Bacon, quickly ensued to find the shortest connection from any actor to Kevin Bacon by actors who had appeared in the same production. Most actors have a very short connection to Kevin Bacon and thus to each other. For example, the shortest route from Charlie Chaplin to Kevin Bacon has length three: Chaplin directed and had a small acting role in the 1967 movie *A Countess from Hong Kong*, with Sophia Loren as the countess. Sophia Loren starred in the little-known 1979 flick *Firepower*. Eli Wallach had a major role in *Firepower* and a tiny uncredited role in *Mystic River* with Kevin Bacon.

Mathematicians have a similar game for having co-written papers centering on the highly prolific combinatorialist Paul Erdős.[*]

The researchers at the Institute first decided to check the six-degrees rule for friendships in Frenemy. To check whether Alice and George have a chain of length six between them, one simple approach is to try all possible chains of length six between them. But there are 3,198,400,279,980,000,480,000 possible chains of length six through the 20,000 inhabitants of Frenemy, which would take more than 100 years even if a computer could check a trillion chains a second. Is there a better way to check the distance between Alice and George?

Yes. There is a simple process that will quickly compute how close Alice and George are to each other.

- Assign Alice the number 0.
- Assign all of Alice's friends the number 1.
- Take all the people who have been assigned number 1, and number all their friends 2 (except for those friends who already have a number).
- Take all the people who have been assigned number 2, and number all their friends 3 (again, except for those friends who already have a number).
- Continue until George gets a number.
- George's number is the distance between Alice and George.

Such informal descriptions of computational procedures are known as *algorithms,* named after Muhammad ibn Mūsā al-Khwārizmī, a ninth-century Persian mathematician. In AD 825, al-Khwārizmī wrote a treatise, *On the Calculation with Hindu Numerals,* that was principally responsible for spreading the Indian system of counting throughout the Middle East and Europe. It was translated into Latin as *Algoritmi de numero Indorum.* The name Al-Khwārizmī, rendered in Latin as Algoritmi, led to the term *algorithm.*

---

[*] I have written papers with three different co-authors of Paul Erdős, giving me an Erdős number of 2. With Erdős's 1996 passing, my chances of reducing my Erdős number are slim. I have had no acting experience (or talent) and do not have a Bacon number.

The algorithm above finds how close Alice and George are to each other in about half a million computation steps. To find the lengths of separations between all pairs of inhabitants of Frenemy we need a more clever procedure, known as the Floyd-Warshall algorithm, which will compute all these distances in about eight trillion computation steps. Now, a trillion sounds like a large number. but even a personal computer these days can do several billion operations a second. The machines at the Frenemy Institute could completely compute the separation distance for all of Frenemy in a couple of minutes. The scientists discovered the average separation in Frenemy was slightly more than six, though there were a few groups of friends that were completely isolated, not friends with anyone else.

Let's not underestimate what just happened. While the researchers could write a computer program to simply check all possible paths to find the shortest path of friendship connecting Alice and George, such a program would have to check so many paths it could not complete its task in any reasonable time. But with some rather simple algorithms they could compute the separation distance between Alice and George in well under a second, and between all pairs of people in Frenemy in a couple of minutes.

Matchmaking

Successful relationships in Frenemy depend almost entirely on a strong friendship relation. But relationships typically form haphazardly, which works well for those who find someone compatible but leaves many unable to find a good partner.

The Frenemy researchers realized they could use their data for helping society by maximizing the number of successful marriages. They put out a call on their web page and quickly got 200 volunteers, an equal number of heterosexual men and women. They wanted to make as many compatible relationships as possible.

How many possible pairs does one have to search through? There are 100 possible women who could potentially match up with the first man. After that choice is made there are 99 women left to match with the second man, 98 to match with the third and so on. This means we have 100 times 99 times 98 times . . . times 2 times 1, a value known as "100 factorial" and written 100!, a 158-digit number. 100! is much larger than a googol, a name invented by a child, the nine-year-old nephew of the mathematician Edward Kasner, who was looking for a name for the number that is 1 followed by 100 zeros.

Google, the Internet company, got its name from a misspelling of googol to give a symbolic but highly inaccurate representation of the large amount of data analyzed by its search engine. As large as the Internet is (which is impossible to actually measure), it doesn't come close to a googol of information at any level of granularity. Certainly, even if we hooked up all the computers in the world over all eternity, there is no hope of searching through a googol, and certainly not 100! possibilities.

Yet the researchers in Frenemy could still find the largest possible number of successful matches. They would just need to use clever algorithms. The diagram below shows the couples who are friends.

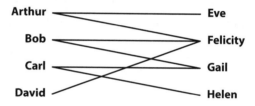

Figure 3-2. Couples in Frenemy.

Now let's see who might be romantic matches. Start with Arthur and match him up with Eve. Bob and Felicity are unmatched friends, so let's match them up, and the same for Carl and Gail. Now we have a diagram, where the dotted lines are the matches we have made.

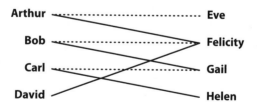

Figure 3-3. Couples Match Partial.

There are no more unmatched friends, so have we achieved the best possible overall matching of people? Not quite.

David is unmatched but is friends with Felicity. Felicity is matched to Bob. Bob is friends with (but not matched to) Gail. Gail is matched to Carl. Carl is friends with the unmatched Helen. If we break up Bob and Felicity and Carl and Gail and rematch them, now everyone has a mate.

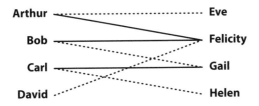

Figure 3-4. Couples Match Full.

The sequence of solid and dotted lines that starts and ends with unmatched people in the second diagram is called an *augmenting path*, and we can always use it to get a larger matching. In 1957, Claude Berge showed that every matching that isn't the largest possible has an augmenting path. The computer scientists at the Frenemy Institute wrote a straightforward algorithm to find these augmenting paths and were able to match up 98 percent of the people in the study.

Shortly afterward the Frenemy Supreme Court ruled that Frenemy must allow marriages between people of the same gender. The Frenemy Institute put a new call on its web page for volunteers of all sexual orientations, leading to much more complicated diagrams, including overlapping love triangles such as that shown below.

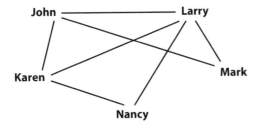

Figure 3-5. Nontraditional Coupling.

The simple algorithm Frenemy scientists had used for finding the augmenting paths no longer worked so well. Instead they turned to the work of Jack Edmonds. In 1965 in a paper beautifully titled "Path, Trees and Flowers," Edmonds developed a bit more complicated method for finding the augmenting paths in general friendship diagrams. The Frenemy Institute implemented Edmonds's ideas and was able to match 97 percent of the people in the second study.

"Path, Trees and Flowers" had a greater impact than just giving an efficient method for solving the general matching problem. Edmonds's algorithm takes approximately 100^4 computation steps to find the best possible matching for a group of 100 people. The number 100^4 is 100 million (100,000,000), not much time at all for today's computers, while the more naive approach of trying all matchings would yield about two quinvigintillion (2 followed by seventy-eight zeros) computation steps. In his paper Edmonds goes into a long digression on the nature of an efficient algorithm. While he realizes that no formal definition can completely capture the intuitive idea of efficiency, he suggests a notion of efficiency by having a procedure that uses computation time that is "algebraic" in the size of the problem, for example, 100^4 or 100^2 or 100^{12}. Later this class of problems would become known as P (for "polynomial," which replaced Edmond's "algebraic") and come to represent problems we can solve efficiently. That's the P side of the P versus NP question.

Cliques

A sociology professor at the Frenemy Institute needed to find fifty inhabitants every pair of which were friends of each other, for a study she wanted to run. Unable to find this group on her own, she discussed the matter with some computer scientists, who described the friendship data they developed. "It shouldn't be a problem to find your fifty-person clique of friends," said one of the CS profs.

But it was a problem—a very difficult problem. The number of different fifty-person groups in Frenemy was an extremely large, 151-digit number; checking even a small fraction of them was out of the question. The researchers tried every trick, realizing, for example, that if a person had fewer than forty-nine friends, that person couldn't be part of a fifty-person clique. But as clever as they were, they couldn't find even twenty-five people who were friends with each other or produce any convincing argument that no fifty-person cliques existed in Frenemy.

Just as they were about to give up, one of the grad students said, "What about the Alpha Society?" The Alpha Society was a legendary, semi-secret society whose members were rumored to all be friends of each other. The computer scientists tracked down fifty members of the Alpha Society (since it was after all only semi-secret). Once they had a list of fifty people there were only 1,225 pairs of friends to check within this group. Surprising the computer scientists, but not the members of the Alpha Society, all of these people were friends of each other. The computer scientists had their clique.

Passing the Rod

Let's see how even a small change can make a problem with a very easily computable solution into a problem where the solution is quite hard to find.

The children of Frenemy have a game called Pass the Rod. A group of kids have a stick they pass around. A "pass" is when two of the kids hold the stick together as it moves from one to the other.

There are two rules:

1. The stick can only be passed between friends.
2. Every pair of friends in the group should pass the stick exactly once between them.

Suppose five kids are playing this game.

The kids can successfully pass the rod by starting with Barbara, who passes the stick to Eric, who passes to Alex, who passes to Cathy, who passes back to Eric, who finally passes the rod to David.

Kids who play the game quickly discover they can only win if at most two people in the group have an odd number of friends in that group: Only David and Barbara have an odd number of friends, one each, while Cathy and Alex each have two and Eric has four. Why? Because other than the first and last person, each person receives the stick as often as he or she gives the stick to one of his or her friends, and hence each of these people handles the stick an even number of times.

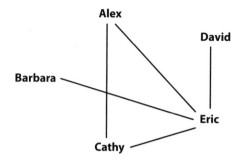

Figure 3-6. Kids.

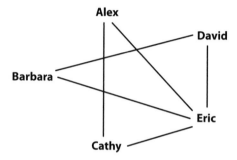

Figure 3-7. Kids Even.

If all the players have an even number of friends, then the kids can still win by starting and ending with the same person.

The kids could win by starting with Alex and having Alex pass the stick to Eric, who passes to David, who passes to Barbara, who passes back to Eric, who passes to Cathy, who finally passes back to Alex.

Pass the Rod comes from a famous puzzle from the eighteenth century. The city of Königsberg in Prussia (now Kaliningrad in Russia) had seven bridges along the branches of the Pregel river, as shown on this old map.

The citizens of Königsberg wondered if anyone could cross every bridge exactly once. In 1735, the famous mathematician Leonhard Euler drew a diagram like this to represent the puzzle.

This is similar to a Pass the Rod game except that now there are multiple friendships between North and the Island and South and the Island. Nevertheless, the same principle applies, and Euler showed that no one could cross each bridge exactly once since all four land masses have an odd number of bridges.

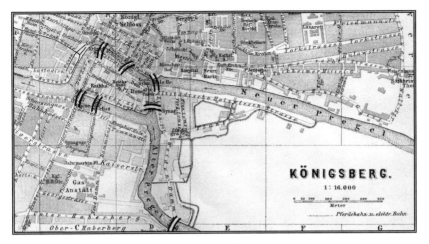

Figure 3-8. Bridges of Königsberg.

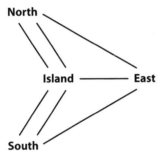

Figure 3-9. Euler Puzzle.

Because of Euler and the Seven Bridges of Königsberg puzzle, any solution to a Pass the Rod game is called an Eulerian path. There are many possible ways to try and play Pass the Rod with a large number of people, but the Frenemy children can easily determine whether a solution exists just by seeing how many people have an odd number of friends. Once they know a solution exists, the kids have little trouble piecing it together. Eulerian paths are another example of a problem in P, one that has efficient solutions.

As the kids of Frenemy get older they tire of Pass the Rod, finding it too easy to solve. So they start playing a variation of the game called, unimaginatively, Pass the Rod II, with the following rules:

1. The stick can only be passed between friends.

2. Every person should hold the stick exactly once, except for the first person, who gets the stick back at the end.

In this friendship diagram David passes the Rod to Barbara, who passes to Eric, who passes to Alex, who passes to Cathy, who passes back to David.

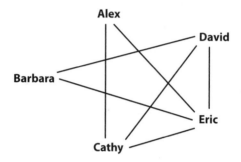

Figure 3-10. Pass the Rod Possible.

In the next friendship diagram, the gang found it impossible to win the Pass the Rod II game.

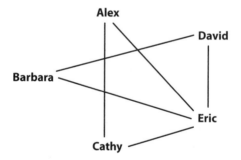

Figure 3-11. Pass the Rod Impossible.

Given the simpler rules, children think that solving a Pass the Rod II game should be even easier than solving Pass the Rod. But once the size of the group gets larger, solving Pass the Rod II seems to get much more difficult. In 1857 the mathematician William Rowan Hamilton invented the Icosian game, which we will describe as a Pass the Rod II puzzle, just using initials of people's names to keep the diagram simpler.

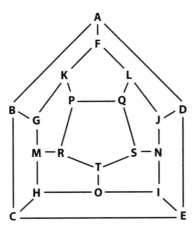

Figure 3-12. Icosian Puzzle.

The Icosian game comes from a regular dodecahedron, a ball with twelve flat faces.

If the vertices are people in Frenemy, where each edge of the dodecahedron indicates a pair of friends, you get the friendship diagram above. Can you find the Pass the Rod II solution and solve the Icosian game? You can find the solution at the end of the chapter.

In honor of the inventor of the Icosian game, Pass the Rod II solutions are called Hamiltonian cycles.

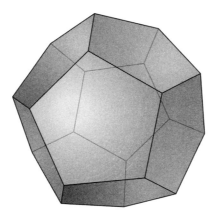

Figure 3-13. Dodecahedron.

Painting Houses

The government of Frenemy passed a new law that required (for aesthetic reasons) that all citizens paint their houses a different color from their neighbors' houses, whether they be friend or foe. After some uproar about forcing citizens to spend additional money on paint and labor, the government agreed to foot the bill for painting the houses, on the condition that the government would get to choose the colors.

Since the government had to invest in large amounts of paint, government functionaries wanted to keep the number of different colors to a minimum. Even using one less color would save millions of dollars. The government gave a grant to the Frenemy Institute to find the smallest number of different colors needed to paint all the houses and make sure no neighboring houses were of the same color.

No one in Frenemy has more than twelve neighbors. So a quick-and-dirty-approach requires only thirteen colors: just color each house differently than all the neighbors' houses that have previously been colored. But the Frenemy researchers could do much better.

In 1852 a South African mathematician, Francis Guthrie, was coloring a map of the counties of England and suspected that four colors would suffice to color any map so that two counties that border have different colors. Guthrie's question was considered by various mathematicians of the time, and two "proofs" that four colors sufficed were announced, by Alfred Kempe in 1879 and by Peter Tait a year later. These proofs lasted eleven years before each was found to have fatal flaws, leaving the problem open for nearly a century more.

In 1976 Kenneth Appel and Wolfgang Haken finally gave a proof that four colors suffice, using a controversial technique. Their proof required using a computer to verify all the cases needed to make their argument work. Traditional mathematicians didn't trust a proof not completely checked by a human mind, but many decades later no mistakes have been found in the Appel-Haken proof, and few today doubt that every map can be four-colored.

Is four the limit? Can any map be colored with three colors? No. Consider Nevada and its neighbors.

California, Oregon, Idaho, Utah, and Arizona form a ring around Nevada. Since this ring has an odd number of states (five), one needs at least

Figure 3-14. Nevada.

three colors to color California, Oregon, Idaho, Utah, and Arizona: If we had only two colors, green and blue, say, and used green for California, then Oregon would have to be blue since it borders California, Idaho would have to be green, and Utah blue. Now, Arizona borders both green-colored California and blue-colored Utah, so it can't be green or blue. So we need three colors, blue, green, and yellow, to color these five states.

Nevada borders all five of these states, so Nevada can't be colored blue, green, or yellow. So we need a fourth color, red, to color Nevada and its neighbors.

There are computationally efficient algorithms for finding how to color a map with four colors based on proofs of the four-color theorem. The Frenemy researchers could then find a way to paint the houses of Frenemy with four colors so that no two neighbors had houses painted the same color. The government pressed the Institute to find a way to use only three different colors of paint. By chance, no property in Frenemy had neighbors who formed an odd-lengthed ring around them, so the researchers could not immediately rule out a way to paint the houses with three different colors.

After trying for some time, the computer scientists at the Frenemy Institute gave up. They admitted to the government their inability to find a way to paint the houses using only three colors. The government went ahead creating four colors to paint the houses of Frenemy. The computer scientists at Frenemy had more difficulties getting grants in the future.

Making Groups

The teachers at Frenemy Elementary School wanted to separate their 500 students into two groups. Friends wanted to remain together, so the teachers wanted to have as few friends as possible end up in different groups. Let's consider the friendships we looked at earlier in the Pass the Rod game.

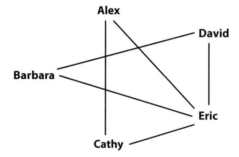

Figure 3-15. Elementary.

The best grouping would put Alex and Cathy in one group and Barbara, David, and Eric in the other, which would cut only two friendships, Alex-Eric and Cathy-Eric. There is no grouping that would break only one friendship.

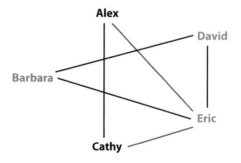

Figure 3-16. Elementary Groups.

The principal of Frenemy Elementary asked the Institute for help. The Frenemy Institute had many efficient algorithms for this problem, known as the min-cut problem because the teachers wanted to keep the number of friendships cut to an absolute minimum. The researchers found a nice splitting of the 500 students into two groups, cutting only seventeen friendships.

And so everyone lived happily ever after. Not quite "ever after" but more like a day before the teachers realized that having many enemies within a group was much worse than limiting the number of split friendships. So the principal went back to the Frenemy Institute and asked the researchers to find two new groups that put as many enemies as possible into different groups. The principal thought that because the Institute had so easily handled the first grouping, it could just as easily solve this new problem. The principal thought wrong.

In this new problem, the researchers needed to make two groups that cut as many enemies as possible, known as the max-cut problem. But unlike with the min-cut problem, computer scientists don't know any efficient algorithms to find the grouping that maximizes the number of enemies that cross between groups.

All was not lost. The Frenemy Institute, using a 1995 algorithm developed by Michel Goemans and David Williamson, found two groups with 1,321 enemy pairs crossing the two groups, and while they couldn't find the best grouping possible they knew no grouping existed with more than 1,505 enemy pairs between the groups. The principal was a bit disappointed that the Institute could not find that best possible grouping, but thought the solution reasonable. And from then on everyone did live more or less happily ever after.

P versus NP

Let's look at the problems where the Frenemy researchers had difficulty finding the best solutions: clique, Hamiltonian path (Pass the Rod II), map coloring, and max-cut. These problems all share a common feature, an easy way to check solutions. Once the Frenemy scientists found members of the Alpha Society, they could easily verify that any two members of the society were friends, and so the Alpha Society forms a clique. If the students succeed in finding a solution to the Pass the Rod II game, they can just play out the game to see if the solution works. Once all the houses are painted the government can in a reasonable amount of time check that each house is painted a different color from its neighbors. And finally, for any separation of students into two groups, the principal can easily count how many pairs of enemies land in the two separate groups.

Computer scientists have a name for problems where we can check solutions quickly: NP (which stands for "nondeterministic polynomial time," if you really need to know). Clique, Hamiltonian path, map coloring, and max-cut are prime examples of problems in NP.

Contrast this with the class P of problems where we can find the best solution quickly: shortest paths, matching, Eulerian paths (Pass the Rod I), and min-cut.

Maybe there is some clever algorithm that would find a clique quickly. Maybe in the future some brilliant graduate student will come up with an easy way to find Hamiltonian paths, an efficient algorithm for coloring maps, or a fast process to group students that maximizes the number of enemies between the groups. Perhaps Clique, Hamiltonian path, map coloring, and max-cut are all in P, just like shortest paths, matching Eulerian paths, and min-cut. Perhaps every problem in NP has solutions we can find quickly. We just don't know.

That's the P versus NP problem. P = NP means we live in the beautiful world where every problem in NP has efficient solutions, every problem for which we have solutions we can check those solutions quickly, and the best solutions can also be found quickly. Conversely, if there is even one problem in NP where no algorithm can find a quick solution, then P ≠ NP.

Determining whether P = NP is the most important question in computer science and perhaps in all of mathematics. Many have tried hard to find good algorithms for problems for Clique, Hamiltonian path, and the others, with little success. Conversely, to show P ≠ NP would require showing that no algorithm, now or in the future, could find cliques or solutions to some other NP problem quickly. How do you show that something is impossible to do? We have not seen much progress in either direction.

The great importance of P versus NP is why the Clay Mathematics Institute is offering a million dollars for its solution. And why I wrote this book.

Beyond Frenemy

This chapter has given only a small sample of the thousands of NP problems for which we don't know how to find solutions quickly. Lest you think that the P versus NP problem is of interest only to computer scientists and the fictional residents of Frenemy, let me mention a small sampling of other NP problems from a few other scientific fields where we don't know efficient algorithms.

Biology

The human genome has twenty-three chromosome pairs. Each of these chromosomes is a sequence of base pairs each of which is described by one of adenine (A), cytosine (C), guanine (G), and thymine (T). The sequence of a chromosome that starts something like ACTGATTACA can be quite long. The shortest sequence has about 47 million base pairs and the longest chromosome has about 247 million. Current DNA sequencing technology can sequence only about 20–1,000 base pairs at a time. Biologists need to take many short sequences and figure out how best to put them together to make the full sequence. Putting these sequences together is a hard computational problem, an NP problem, because if we actually knew the DNA sequence in advance, we could relatively quickly check that it's a good match to the sequences we find. Since no one knows efficient algorithms for solving this NP problem optimally, biologists need to take more sequences to map a genome, and likely that map will have more gaps and mistakes than if they had optimal algorithms for splicing sequences together.

These DNA sequences encode messenger RNA (mRNA) sequences, which in turn describe proteins, which perform critical operations in nearly every function of the cells that make up every living creature. Proteins usually need to take on a specific shape to perform these functions, and via some process an mRNA sequence will cause a protein to fold into a specific shape. This computational process of protein folding remains one of the great mysteries in biology, and the P versus NP problem could have strong implications for understanding protein folding to prevent and fight disease.

Protein threading is a statistical approach that could be used to make some predictions of the protein folds based on the mRNA sequence. Even this limited approach to understanding folding involves solving difficult NP problems.

Physics

Finding the minimum energy state of a physical system such as interactive magnetic particles or soap bubble formation is an NP problem. We don't know how to efficiently find these minimum energy states. Don't physical systems always end up in their lowest energy state? Not always.

Consider a marble placed on the curve in figure 3-17. This marble will have minimum potential energy if it ends up at 3.0. But if the marble is placed at 1.0, it will stay there unless we give it a push. A physical system doesn't always end up in the lowest energy state. Finding a minimum in complex systems can be hard for both computers and physical systems.

Quantum mechanics might help us solve some but not all difficult NP problems. We'll explore this more in chapter 9.

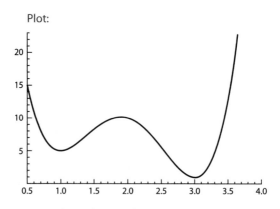

Plot $7x^4 - 55x^3 + 148x^2 - 159x + 64$ from 0.5 to 4.0

Figure 3-17. Physical System Graph.

Economics

A hedge fund manager chooses among many complex investment vehicles. A consumer walks into a supermarket with a limited budget. Both need to solve difficult NP computational problems. Both will often make suboptimal decisions because they can't always solve these problems. How exactly do these computational inefficiencies in markets affect our economy and society in general? That's a great question to which we don't have great answers.

John Nash was a mathematician whose life is explored in the book and movie *A Beautiful Mind*. Nash received (belatedly) a Nobel Prize for showing that strategic interactions among individuals have an equilibrium state, whereby everyone has a strategy such that, if played together, no one gains any advantage from deviating from his or her chosen strategy. Nash's proof does not give a method for finding these strategies, and computer scientists now have evidence that finding such strategies can be a computationally difficult problem. These difficulties suggest that markets will not, on their own, always find these equilibriums, meaning they will always remain in flux as people will continually change their strategies seeking better outcomes.

Mathematics

In 1928 the renowned German mathematician David Hilbert put forth his great *Entscheidungsproblem*, a challenge to find an algorithmic procedure to find the truth or falsehood of any mathematical statement. In 1931 Kurt Gödel showed there must be some statements that any given set of axioms could not prove true or false at all. Influenced by this work, a few years later Alonzo Church and Alan Turing independently showed that no such algorithm exists.

What if we restrict ourselves to relatively short proofs, say, of the kind that could fit in a short book? We can solve this computationally by looking at all possible short proofs for some mathematical statement. This is an NP question since we can recognize a good proof when we see one but finding one in practice remains difficult. This is why mathematicians derive fame from finding clever ways to prove mathematical statements.

It can even be difficult to find ways to make simple logical expressions true. From that problem came a theory that links most of the NP problems together, a story we tell in the next chapter.

A Solution to the Icosian Game

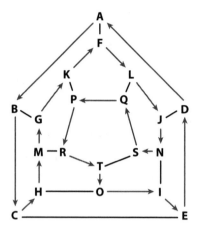

Figure 3-18. Icosian Solution.

— — — — — — — — — —

Chapter 4

THE HARDEST PROBLEMS IN NP

A psychologist decides to run an experiment with a mathematician. The psychologist puts the mathematician in a one-room wooden hut that has some kindling on the floor, a table, and a bucket of water on the table. The psychologist lights the kindling on fire. The mathematician takes the bucket and dumps the water on the kindling, extinguishing the fire.

So far, so good, so the psychologist repeats the experiment with the mathematician in the same hut with the same table, kindling on the floor, and bucket of water, but this time the bucket of water sits on the floor next to the kindling. The psychologist again lights the kindling. The mathematician takes the bucket and puts it on the table. As the hut burns down, the psychologist and her colleagues pull the mathematician out of the hut just in time.

The psychologist asks the mathematician why he didn't just put out the fire as before. The mathematician responded, "I just reduced the problem to a previously solved case."

—Old math joke

The First NP-Complete Problem

Tom Hull, chair of the University of Toronto Computer Science Department in 1970, wanted to hire Steve Cook. Steve Cook had just been denied tenure at the University of California at Berkeley. Cook enjoyed sailing, so Hull took him out on Lake Ontario to show that sailing near Toronto was just as good as sailing in San Francisco Bay. The ploy worked, and Steve Cook joined the University of Toronto faculty in the fall of 1970. A brilliant move, since Cook would soon become Canada's most famous computer scientist.

Cook studied the connections between logic and computer science. That fall he submitted a paper for consideration to the Third ACM Symposium on the Theory of Computing (STOC), to be held the following May. The paper he submitted had some results in it that had long been forgotten, but it raised enough interest for the paper to be accepted for presentation at the conference. For the conference proceedings Cook rewrote the paper with new work, and that paper, "The Complexity of Theorem-Proving Procedures," would bring us the P versus NP problem and change history.

To understand Cook's paper, let's go back to the clique problem from the last chapter. Remember that a clique is a group of people in Frenemy all of whom are friends with each other. In this friendship diagram Alex, Cathy, and Eric are a clique but Alex, David, and Eric are not since Alex and David are enemies and not friends.

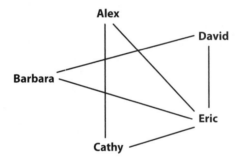

Figure 4-1. Cliques.

In Frenemy, as you might recall, there is a semi-secret society, the Alpha Society, a large group of Frenemies that claim they are all friends with each other, forming a large clique. What can we glean about the Alpha Society from the friendship diagram above, if they are indeed a clique? We can't rule out any single person from being in the Alpha Society. Alex could be in the Alpha Society and David could be in the Alpha Society. But Alex and David can't be both in the Alpha Society since they are enemies. So either Alex is not in the Alpha Society or David isn't. Let's write that out logically.

Alex not in Alpha Society OR David not in Alpha Society

OR here is not exclusive; it is possible that neither Alex nor David is in the Alpha Society. Since Alex and Barbara are enemies we also know they both can't be in the Alpha Society:

Alex not in Alpha Society OR Barbara not in Alpha Society

Both of these logical expressions have to be true, so we have:

(Alex not in Alpha Society OR David not in Alpha Society) AND
(Alex not in Alpha Society OR Barbara not in Alpha Society)

Since Barbara and David are friends they could both be in the Alpha Society, and this is not ruled out by this logical expression. Finishing off this expression for the full diagram we get:

(Alex not in Alpha Society OR David not in Alpha Society) AND
(Alex not in Alpha Society OR Barbara not in Alpha Society) AND
(David not in Alpha Society OR Cathy not in Alpha Society) AND
(Cathy not in Alpha Society OR Barbara not in Alpha Society)

If Alex, Cathy, and Eric were in the Alpha Society but Barbara and David weren't, the expression would be true: each of the OR terms has someone not in the society. If instead we look at Alex, David, and Eric, if they were in the society the expression would be false: the term (Alex not in Alpha Society OR David not in Alpha Society) would be false since both Alex and David are in the society.

This expression nicely captures cliques. It is true exactly when the members of the Alpha Society form a clique.

We can create a similar logical expression for all of the 20,000 inhabitants of Frenemy, which we'll call Φ for short. Φ could be quite large, perhaps several million characters long, but not so large a computer couldn't store it easily. The Alpha Society would make Φ true if it indeed were an actual clique.

Let's have another logical expression, $\Psi 50$, that will be true if the Alpha Society has at least fifty members. I won't go into the details on the exact formulation of $\Psi 50$, but we can build such an expression based on how the very first digital computers added numbers.

If we combine our two expressions as ($\Psi 50$ AND Φ) we now have a single expression that will be true if the Alpha Society is a clique with at least fifty members. Conversely, if the Alpha Society makes ($\Psi50$ AND Φ) true, then the Alpha Society is a clique of size at least fifty.

A logical expression is satisfiable if there is some way to put people in and out of the Alpha Society to make the expression true. The expression ($\Psi50$ AND Φ) is satisfiable exactly if there is some society that is a clique with at least fifty people in it.

Suppose we had some quick algorithm that told us whether or not a logical expression is satisfiable. If we feed ($\Psi50$ AND Φ) into the algorithm and it says yes, ($\Psi50$ AND Φ) is satisfiable, then Frenemy does have a clique of fifty people. If ($\Psi50$ AND Φ) is not satisfiable, then there can be no such clique. Solve the satisfiability problem and you can solve the clique problem.

We just described one of the most important concepts in computer science: a reduction. We reduced the problem of finding cliques to the problem of checking logical satisfiability. Any method we have for solving the satisfiability problem can now be easily converted to an algorithm for the clique problem. If satisfiability has efficient algorithms, then so does the clique problem. Clique is at least as easy to solve as satisfiability. If satisfiability is easy to solve, then so is clique. If clique has no efficient algorithms, then satisfiability doesn't either.

Not only does clique reduce to Satisfiability, but so do the other NP problems we discussed earlier, including the traveling salesman problem, the Hamiltonian path, max-cut, and map coloring. What Steve Cook did was show that every problem in NP has a reduction to satisfiability. Solve satisfiability and you can solve all of NP. If you have an efficient algorithm for solving satisfiability, then all the problems whose solutions we can efficiently check have efficient algorithms, and P = NP. You just need one efficient algorithm for satisfiability and the Clay Institute's million dollars is yours.

Satisfiability is itself an NP problem since we can efficiently check whether a specific setting makes the expression true, making satisfiability the hardest of all problems in NP. Determining whether or not satisfiability has efficient algorithms is the same as determining whether P = NP.

Steve Cook presented his paper at the STOC conference on May 4, 1971, in what was then the Stouffer's Somerset Inn in Shaker Heights, Ohio.

The result suggests that Satisfiability is a good candidate for an interesting problem not efficiently computable, and I feel it is worth spending considerable effort trying to prove this conjecture. Such a proof would be a major breakthrough.

And thus the P versus NP problem was born.

Twenty-One

While viewed as an important result, Cook's paper was not an immediate game-changer. Satisfiability was a problem of limited interest, and NP was not yet well understood. Steve Cook didn't even give NP a name in his paper, instead using just its namesake, "nondeterministic polynomial time." But that would change soon with a follow-up paper by Berkeley professor Richard Karp.

Soon after seeing Cook's paper, Karp realized how to reduce satisfiability to clique. Given a logical expression, Karp created a simple procedure to convert that logical expression to a friendship diagram where the logical expression would be satisfiable exactly when the friendship diagram had a clique of a certain size. Any procedure that solved clique could be used to solve satisfiability. Cook showed that satisfiability is the hardest problem among all of NP problems. Karp shows that clique is at least as hard as satisfiability. That makes clique, too, one of the hardest problems in NP. Like satisfiability, if we had some quick algorithm to solve clique, then every problem in NP would have efficient algorithms and P = NP.

Cook showed how to reduce problems like clique to satisfiability. Karp showed how to reduce satisfiability to clique. The very different-looking problems of satisfiability and clique are computationally the same. Satisfiability is easy to compute if and only if clique is easy to compute if and only if P=NP.

Karp showed not only that clique was one of the hardest problems in NP, he also showed that nineteen other important problems were also among these hardest, including the partition puzzle, traveling salesman,

Hamiltonian path, map coloring, and max-cut. Solve any of these problems efficiently and you have an efficient algorithm for all of them and have shown P = NP. If P ≠ NP, then none of these twenty-one problems (including satisfiability) has a quick algorithm.

Karp did not invent these twenty-one problems, nor were these problems solely of interest to mathematicians and computer scientists. Many of these problems came from the real world.

Consider Coca-Cola, with its three thousand-plus brands of beverages sold worldwide. A single Coca-Cola bottling plant might produce a few hundred different drink products. Each plant has several machines, and each of these machines runs a series of jobs that mixes some of the materials needed for each beverage. These machines can produce different products, and the bottling plants want to schedule the jobs on various machines to get the best throughput, producing the most amount of the desired beverages in the least amount of time. This is the job scheduling problem, another of the twenty-one that Karp showed as hard as any other NP problem.

Since the development of the digital computer, scientists and programmers have tried their best to develop the best algorithms for the job scheduling problem because a schedule that results in even a small productivity improvement could save a company millions of dollars. No one has managed to create algorithms that always give the best job schedule. Karp showed that even simple variations of job scheduling are as hard as any NP problem, giving an instant explanation for why people can't find those algorithms.

But it's not just big corporations. Suppose you take the family to Disney World during spring break when the lines are long. You want to see as many of the best attractions as you can without spending much time in line. The authors of the *Unofficial Guide to Disney World* developed a series of tours that try to minimize the wait, but they acknowledge the difficulty of their task.

> The 21 attractions in the Magic Kingdom One-Day Touring Plan for Adults has a staggering 51,090,942,171,709,440,000 possible touring plans. That's over 51 billion billion combinations, or roughly six times as many as the estimated number of grains of sand on Earth. Adding in complexity such as FASTPASS, parades, meals and breaks further increases the combinations.

Scientists have been working on similar problems for years. Companies that deliver packages, for example, plan each driver's route to minimize the distance driven, saving time and fuel. In fact, finding ways to visit many places with minimal effort is such a common problem that it has its own nickname: the Traveling Salesman problem.

Many scientists have tried to find the best solution for the traveling salesman problem. Others have tried to find great algorithms for the job scheduling problem. Others have tried for clique, max-cut, and most of the rest of Karp's twenty-one problems. Karp's paper meant that all these researchers were in fact working on the same problem, for if any of them had succeeded in finding an efficient procedure for one problem, it would have given an efficient procedure to all of them.

Likewise, all these people working independently on all these different problems have failed to find good algorithms for them. That gives some indication that P ≠ NP, or at least it would be very difficult to find a good algorithm for any of these problems. The jobs scheduling people can go tell their bosses that they couldn't find the best way to schedule the jobs on their machine but even those hackers in Orlando can't figure their way around Disney World.

In one fell swoop, Karp tied together all these famous difficult-to-solve computational problems. From that point on, the P versus NP question took center stage.

Every year the Association for Computing Machinery awards the ACM Turing Award, the computer science equivalent of the Nobel Prize, named for Alan Turing, who gave computer science its foundations in the 1930s. In 1982 the ACM presented the Turing Award to Stephen Cook for his work formulating the P versus NP problem. But one Turing Award for the P versus NP problem is not enough, and in 1985 Richard Karp received the award for his work on algorithms, most notably for the twenty-one NP-complete problems.

What's in a Name?

Karp's paper gave the names P and NP that we use today. But what should people call those hardest problems in NP? Cook called them by a technical name "deg({DNF tautologies})," and Karp used the term "(polynomial) complete." But these names didn't feel right.

Donald Knuth took up this cause. In 1974 Knuth received the Turing Award for his research and his monumental three-volume series, *The Art of Computer Programming*. For the fourth volume, Knuth, realizing the incredible importance of the P versus NP problem, wanted to settle the naming issue for the hardest sets in NP. In 1973, Knuth ran a poll via postal mail. He famously doesn't use email today, but back in 1973 neither did anyone else.

Knuth suggested "Herculean," "formidable," and "arduous," none of which fared well in the poll. Knuth got many write-in suggestions, some straightforward, like "intractable" and "obstinate," and others a bit more tongue-in-cheek, like "hard-boiled" (in honor of Cook) and "hard-ass" ("hard as satisfiability").

The winning write-in vote was for "NP-complete," which was proposed by several people at Bell Labs in New Jersey after considerable discussion among the researchers there. The word "complete" comes from mathematical logic, where a set of facts is complete if it can explain all the true statements in some logical system. Analogously, "NP-complete" means those problems in NP powerful enough that they can be used to solve any other problem in NP.

Knuth was not entirely happy with this choice but was willing to live with it. He truly wanted a single English word that captured the intuitive meaning of hard search problems, a term for the masses. In a 1974 wrap-up of his survey, Knuth wrote "NP-complete actually smacks of being a little too technical for a mass audience, but it's not so bad as to be unusable."

"NP-complete" quickly became the standard terminology. It took Donald Knuth about four decades to finish volume 4.

Knuth should have pushed a bit harder for less technical names for "NP-complete," and perhaps for "P" and "NP" as well. The P versus NP problem has taken on an importance that goes well beyond computer science, and using terminology that just abbreviates a technical definition hides this import from outsiders. But terminology gets embedded in the culture over the decades, and at this point it would be difficult to change, even if we had great alternatives.

Knuth also realized that all this fuss to come up with a name would be wasted if in fact P = NP, since then the NP-complete problems would be just the ones in P. But Knuth was "willing to risk such an embarrassment, and in fact I'm willing to give a prize of one live turkey to the first person who proves that P = NP." So prove P = NP and you will win a million dollars *and* a turkey.

Beyond Karp

After Karp's paper, an industry in computer science arose showing the NP-completeness of a large variety of search problems. Over the next several years, many professors and grad students took a number of known search problems (as well as some new ones) and showed them to be NP-complete. A classic 1979 book[*] lists over three hundred major NP-complete problems. NP-complete problems continually arise in computer science, physics, biology, economics, and many other areas that reach that pinnacle of hardness. A Google Scholar search on "NP-Complete" returns over 138,000 scientific articles from 1972 to 2011, nearly 10,000 in 2011 alone. We can't list all of them here, but let me give you a flavor of some of them.

Dominating Set

Are there fifty people in Frenemy such that everyone else is friends with at least one of them? NP-complete.

Partition into Triangles

The dorm rooms of the Frenemy Institute can each hold three students. Can we assign the students to dorm rooms such that no enemies end up in the same room? NP-complete.

---

[*] *Computers and Intractability: A Guide to the Theory of NP-Completeness*, by Michael Garey and David Johnson (New York: W. H. Freeman, 1979).

Large Sudoku Games

Sudoku is a Japanese puzzle with numbers in a 9 × 9 grid as follows.

5	3			7				
6			1	9	5			
	9	8					6	
8				6				3
4			8		3			1
7				2				6
	6					2	8	
			4	1	9			5
				8			7	9

Figure 4-2. Sudoku.

The goal of Sudoku is to fill in the remaining numbers such that every row, column, and dark 3 × 3 square each contains exactly one of the numbers 1 through 9 as such.

5	3	4	6	7	8	9	1	2
6	7	2	1	9	5	3	4	8
1	9	8	3	4	2	5	6	7
8	5	9	7	6	1	4	2	3
4	2	6	8	5	3	7	9	1
7	1	3	9	2	4	8	5	6
9	6	1	5	3	7	2	8	4
2	8	7	4	1	9	6	3	5
3	4	5	2	8	6	1	7	9

Figure 4-3. Sudoku Solution.

Sudoku is an NP problem since it is easy to check a solution. How hard is it to find a solution? Not bad. With a simple backtracking algorithm, a traditional computer can solve a Sudoku puzzle in a couple of seconds.

	L			U			K					G				C	Q					I		X
			Q	H		R		K							V				A	M	T			
D	A	B	H	I			C			X	T				F								V	K
U	V	X	W		D	J	E			I	R	A			O			C	H					
K		G		X	F		B	W	Q	D			L								O			
			Q	I	U				O		S						R						P	N
		E			D	V	K		J		P	Q				L		A	M	I	Y			H
	F		C		R	A				N	U	G				I	W	S			B			
	I		Q	H		O	Y			L		D			B		K	T		U				
	M	G			W	C				T							J		R		D	V		
M		R		E	B					D			C				H		A			G	W	
		P	W			G		A	Y			E					X		N					
		K	Y			L				W	U	T		N	D									
H	L	T	S				W			V		K	X			F				Q	J			
N	B				H		S	Y	F	P		C	I	K		E		L					T	O
L	Q					E		U	R		F				B	I			X	D		J		T
B			A				C					Y	S			U	V	P			X			
T		X	P		J					Q	A				W		E	R	Y		C			
	H			N	Y	Q				X	I	S	E		F			T			K	W	A	
	K	Y	F	T	A			G			P	N				J	O	Q	L					U
V	W				U		P			H			R	G	X					N	M			Q
	G		O			T			F	X		B	N	M			K	C			E			Y
C	U		J		G	Y	N	O	S			I			V		F			B				
I				R	E			W	S	O		J			A					K				
	P			T	C		X	M	D			Q					Y			U	L	O		

Figure 4-4. Large Sudoku.

But what happens if we have a larger puzzle, like this 25 × 25 variant, where every row, column, or square must have every letter from A to Y exactly once?

Now a computer search will start to take significant time, and 100 × 100 games can defeat today's fastest machines.

Large Sudoku games are NP-complete. Think you are good at Sudoku? If you can reliably solve large puzzles, then you can solve satisfiability, traveling salesman problems, and thousands of other NP-complete problems.

Sudoku is hardly the only NP-complete solitaire game. Consider the Minesweeper game built into Microsoft Windows.

Figure 4-5. Minesweeper.

Each number in a square tells how many mines are within one square (horizontally, vertically, or diagonally) from that square. Click on a square and you will reveal that number. Set a flag if you think there are bombs there. Click on a bomb and you lose the game. Determining whether one can win a large Minesweeper game is also NP-complete. Here are the remaining bombs in the above game.

Figure 4-6. Minesweeper Bombs.

Figure 4-7. Tetris.

Then, of course, we have Tetris, where the player slides and rotates pieces to fill a row. Once a row is filled it disappears, and the goal is to play as long as possible until you run out of rows.

Pieces come in different shapes.

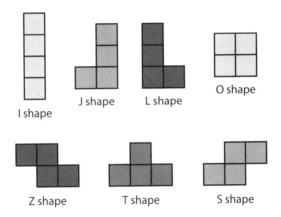

Figure 4-8. Tetris Pieces.

In traditional Tetris you don't know what shape will come next. But even if you knew the order of shapes ahead of time, playing Tetris optimally is once again NP-complete.

Who would think playing Sudoku, Minesweeper, or Tetris well would show P = NP and solve one of the biggest challenges of our generation?

Figure 4-9. Rubik's Cubes. Photo by Tom van der Zanden

How about Rubik's Cube? Even the $3 \times 3 \times 3$ cube takes a while to learn to solve; imagine how much harder solving larger cubes should be.

Actually not. We have efficient algorithms to solve even large Rubik's Cubes puzzles using a branch of mathematics known as *group theory*. These algorithms don't find the absolutely shortest solution, but they always find reasonably short ways to solve the cube from any starting position that can lead to a solution.

It's surprising how easy Rubik's Cube is to solve, while Tetris, Minesweeper, and Sudoku are hard.

How about two-person games like chess, checkers, Othello, and Go? The large versions of these games are as hard as satisfiability and the rest of the NP-complete problems. But these two-person games are not known to be in NP. I could tell you that White has a guaranteed win with Pawn to King Three, but there is no way for you to verify whether I am telling the truth. Computer scientists generally believe that chess, checkers, Othello, and Go are all much harder than any NP problem.

Kidney Exchanges

Most people have two healthy kidneys that filter out the waste in our systems. If one of them fails the other can still function, allowing the person to live a normal life. Occasionally both kidneys fail, requiring expensive and time-consuming dialysis and a greater risk of death.

A person with two healthy kidneys can donate one kidney to someone who has no healthy kidneys if the kidney is compatible with the new body, which can be checked with a simple blood test.

Suppose Alice's kidneys fail. Alice's husband, Bob, agrees to donate one of his kidneys. If Bob's kidney is compatible with Alice, then a surgeon can remove a kidney from Bob and use it to replace one of Alice's failed kidneys.

But Bob might not have a compatible kidney. Still, we can try a kidney exchange.

Suppose Charlie needs a kidney and his brother, David, is willing to donate, but David's kidney is not compatible with Charlie. However, if David's kidney is compatible with Alice and Bob's kidney is compatible with Charlie, they can undergo a simultaneous four-person operation, and everyone ends up with a working kidney.

Suppose we had a kidney database with all the potential donor and recipient couples. We could run efficient algorithms to find the largest number of possible pairwise kidney exchanges. This is basically the matchmaking problem from the last chapter and easy to solve.

No need to stop at two couples. In late 2011, sixty surgeries helped give thirty people healthy kidneys they wouldn't have been able to get in any other way.

If we allow for more than two exchanges, finding the largest number of people we can help is now NP-complete. Here P = NP could save lives. This seems a bit more important than playing a good game of Minesweeper.

The Ones That Got Away

Most of the NP problems that people considered in the mid-1970s either turned out to be NP-complete or people found efficient algorithms putting them in P. But some NP problems refused to be so nicely and quickly characterized. Some would be settled years later, and others we still don't know.

Graph Isomorphism

Several hundred people in Frenemy loved to play Blade Quest, a popular MMORPG (massively multiplayer online role-playing game). As in most other MMORPGs, in Blade Quest each person has a new identity, an avatar, a character they play that interacts with other characters played by other Frenemies. Friendships carry into the Blade Quest world. The avatars of two friends in Frenemy will be friends in Blade Quest. The avatars of two enemies in Frenemy will also be enemies in Blade Quest.

Isabel, John, Kevin, Laura, Molly, and Nancy are a group of people in Frenemy who like to play Blade Quest.

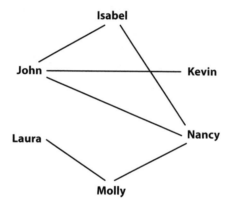

Figure 4-10. Blade Quest.

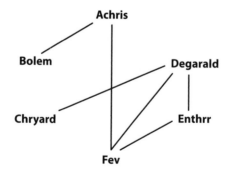

Figure 4-11. Blade Quest Avatars.

This group of six has avatars Achris, Bolem, Chyrard, Degarald, Enthrr, and Fev in the Blade Quest world, though which real person is using which avatar is supposed to be a closely kept secret. Inside Blade Quest these avatars have a friendship diagram.

Laura looked at the two diagrams and had her avatar send a message to the other avatars: "I know who you really are." Can you figure it out?

The only way to match the two diagrams is if Isabelle's avatar is Enthrr, John's avatar is Degarald, Kevin is Chryard, Laura is Bolem, Molly is Achris, and Nancy is Fev. For example, Molly's friends in Frenemy are Laura and Nancy, corresponding to Achris's friends in Blade Quest, Bolem and Fev.

The problem of matching friendship diagrams in this way is known as graph isomorphism. Depending on the diagrams there might be more than one way to match them or no way at all. The graph isomorphism problem is in NP: if you know who is matched to whom you can easily check each pair of people to see that the friendship relationships stay the same.

But whether graph isomorphism is in P, whether we have some efficient algorithm that can always find a matching when one exists, is still unknown. Nor do we know if graph isomorphism is NP-complete and for various technical reasons, computer scientists don't believe that it is. Graph isomorphism is one of the few problems whose difficulty seems somewhat harder than P but not as hard as NP-complete problems like Hamiltonian paths and max-cut.

Prime Numbers and Factoring

You can break down (factor) the number 15 as 15×1 or 5×3. The number 24 is equal to 24×1 or 12×2 or 8×3 or 6×4. But you can only write the number 17 as 17×1. 17 is a prime number, a number you can write as a product only of 1 and the number itself. The first few primes are 2, 3, 5, 7, 11, 13, 17, 19. There is an infinite number of prime numbers. The largest known prime (as I write this) is a number with 12,978,189 digits that begins 316,470,269,330,255,923,143,453,723, . . .

How can you tell whether a number is prime? To check that a number like 1,123,467,619 is prime, you could check all numbers up to

1,123,467,619 and see if any of them divide into 1,123,467,619. Actually, you only need to check the possible factors up to 33,518 (the square root of 1,123,467,619). Sounds quick enough, but how do you check whether

8,273,820,869,309,776,799,973,961,823,304,474,636,656,020,157,784,
206,028,082,108,433,763,964,611,304,313,040,029,095,633,352,319,623

is prime?

Algorithms for checking prime numbers date back to the ancient Greeks, but not until the 1970s did we see methods that worked on numbers with hundreds of digits. These methods relied on using randomly generated numbers in their tests using techniques drawn from an area of mathematics called *number theory*. While these tests worked well in practice, it wasn't until 2002 that an Indian professor, Manindra Agrawal, working with two students, Neeraj Kayal and Nitin Saxena, finally found an efficient algorithm for prime testing that did not need to use random numbers, putting primes in P.

These algorithms would tell us that

8,273,820,869,309,776,799,973,961,823,304,474,636,656,020,157,784,
206,028,082,108,433,763,964,611,304,313,040,029,095,633,352,319,623

is not prime, but, rather surprisingly, they don't tell you the actual factors, the numbers that divide into

8,273,820,869,309,776,799,973,961,823,304,474,636,656,020,157,784,
206,028,082,108,433,763,964,611,304,313,040,029,095,633,352,319,623.

While we can test whether a number is prime efficiently, we don't know any efficient algorithms for actually finding the factors.

Factoring is an NP problem. If you see the numbers

84,578,657,802,148,566,360,817,045,728,307,604,764,590,139,606,051

and

97,823,979,291,139,750,018,446,800,724,120,182,692,777,022,032,973,

you can multiply them together quickly and see that the product is

8,273,820,869,309,776,799,973,961,823,304,474,636,656,020,157,784,
206,028,082,108,433,763,964,611,304,313,040,029,095,633,352,319,623.

Computer scientists don't believe that factoring is in P, nor do we believe that factoring is NP-complete. Factoring is a difficult problem, but maybe not as hard as satisfiability or map coloring.

The importance of primes and factoring goes way beyond mathematicians loving their numbers. Much of modern cryptography uses numbers that seem hard to factor, the topic of chapter 8.

Linear Programming

Frenemy Fancy Franks sells four varieties of sausages: frankfurters, Italian, bratwurst, and chorizo. Each kind of sausage has various ingredients with different costs. Different sausages take different times to manufacture, and they sell at different prices. How much of each sausage should Frenemy Fancy Franks make in order to maximize profits?

Finding the best allocation of sausages means maximizing revenue while fulfilling various requirements. Suppose frankfurter meat costs $1 per sausage, Italian sausage meat is $2 per sausage, $3 for bratwurst, $4 for chorizos, and there is a total meat budget of $10,000 per day. Then one times the number of frankfurters plus two times the number of sausages plus three times the number of bratwursts plus four times the number of chorizos must be at most 10,000.

Optimizing your choices given these kinds of requirements is a problem known as linear programming. The set of possible solutions forms what is known as a polytope in high-dimensional space.

Back in 1947, Georg Dantzig created the simplex method, which solves the linear programming problem pretty quickly. The simplex algorithm takes a walk along the edges of the polytope until it finds the optimal solutions.

Given the simplex algorithm, why do I put the linear programming problem in this category? Because the simplex algorithm may not solve linear programming quickly in some rare instances.

In 1979, Leonid Khachiyan created the ellipsoid algorithm, which chops up the polytope into smaller and smaller pieces until it narrows

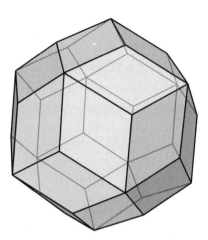

Figure 4-12. Polytope.

it down to the optimal solution. Kahachiyan gives a proof of efficiency of the ellipsoid algorithm, putting linear programming in P. However, the ellipsoid algorithm takes much longer than simplex in practice. The ellipsoid algorithm did inspire many other more complex algorithms in the decades to come, making the ellipsoid algorithm an extremely influential one.

So the linear programming problem has good algorithms in theory and practice—they just happen to be two very different algorithms.

In 1984, Narendra Karmarkar developed the interior point algorithm. Karmarkar's algorithm takes a walk like the simplex algorithm but through the polytope instead of around it. Like the ellipsoid algorithm, interior point is known to be in P, and after some optimization the algorithm performs favorably compared with simplex.

That's three very different algorithms for solving the linear programming problem. The first (simplex) works well in practice. The second (ellipsoid) works well in theory. The third (interior point) works well both in theory and practice. Not bad for a problem still considered unresolved into the late 1970s.

Chapter 5

— — — — — — — — —

THE PREHISTORY OF P VERSUS NP

One does not fear the Perebor, but rather uses it reasonably.[*]

IN THE LAST CHAPTER WE RECOUNTED Donald Knuth's ultimately unsuccessful attempt to find a good English word to capture NP-completeness. Knuth could have turned east to the Russians to find perebor (Перебор). *Perebor* means "brute force search," the process of trying all possibilities to find the best solution. P versus NP asks whether we need perebor to solve the clique problem or whether some faster approach could work.

But Knuth and others in America couldn't so easily look toward Russia. An Iron Curtain separated Russia and Eastern Europe from the United States and Western Europe starting at the end of World War II in 1945. The Cold War created a competition between the United States and Russia that led to a great emphasis on scientific development from the 1950s onward in an attempt to win an intellectual arms race. It also had the negative effect of severely limiting scientific travel and communication between the East and West. While these boundaries started to open up in the 1970s, it wasn't until the end of the Cold War in 1991 that full-fledged research discussions developed between the two countries.

---

[*] Georgy M. Adelson-Velsky, Vladimir Arlazarov, and Mikhail Donskoy, *Algorithms for Games* (*Programirovanie Igr*) (New York: Springer-Verlag, 1988).

Today, with most academic work available over the Internet and with generally open travel around the world, we can now view ourselves as one large research community instead of two separate ones.

This chapter tells two tales, two separate paths that led to the P versus NP question. In the end it was Steve Cook in the West and Leonid Levin in the East who would first ask whether P = NP. Science doesn't happen in a vacuum, and both sides have a long history leading to the work of Cook and Levin. In this chapter we cover just a small part of those research agendas, the struggle in the West to understand efficient computation and the struggle in the East to understand the necessity of perebor. Both would lead us to P versus NP.

The West

The search for efficient algorithms goes back 3,000 years or so when humans first started using basic arithmetic to add large numbers quickly. Our story, though, begins in the 1930s, when humans started to create a theory of algorithmic processes.

Alan Turing

We have explored space and developed telescopes that let us see deep into the distant parts of the universe and its early history. We have built microscopes that let us see atoms, and we have built big machines that smash little particles together so we can find even littler particles. We have decoded the human DNA. But one of the greatest mysteries lies in that device that we have on our desks, in our cars, even in our pockets: something we call a computer. But what is a computer?

The word *computer* dates back to the seventeenth century, well before anyone imagined mechanical devices doing calculation. A computer was a human who computed, did calculations. The growing banking industry relied on these human computers to keep track of deposits and loans.

According to the *Oxford English Dictionary*, the first use of the word *computer* as a mechanical device dates to 1897 and the first use as a digital computer came in the 1940s. The computer has gone from a person to a few big machines to something we use all the time.

Let's not think about computers as actual devices but in terms of what they do: take in some information, process it according to a set of instructions, and produce some outcome. The postal service is a computer: it takes your letters, decodes the address information, and routes them to the proper location. Biology is a computer: it takes a DNA sequence to produce proteins that perform the necessary functions that make life possible.

What about the process we call computation? Is there anything we can't compute? That mystery was solved before we even had digital computers by the great mathematician Alan Turing in 1936. Turing wondered how mathematicians thought, and came up with a formal mathematical model of that thought process, a model we now call the Turing machine, which has become the standard model of computation.

Alan Turing was born in London in 1912. In the early 1930s he attended King's College, Cambridge University, where he excelled in mathematics. During that time he thought of computation in terms of himself as a mathematician. A mathematician has a process in mind, a limited amount of memory, and a large supply of paper and pencils. The mathematician would scribble some notes on a page and either move on to the next page or go back to the previous one where he could modify what he wrote before. Turing used this intuition to create a formal model of computation we now call the Turing machine.

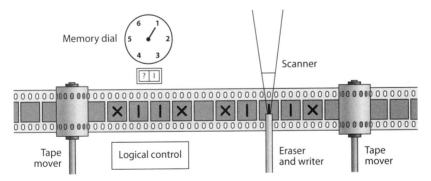

Figure 5-1. Turing Machine.

Though the machine was a very simple model, Turing stated that everything computable was computable by his machine. Alonzo Church at about the same time made a similar claim based on his Lambda Calculus, a rudimentary programming language. This Church-Turing thesis, developed before the invention of modern computers, has stood the test of time. Everything computable, now or in the future, is indeed computable by a Turing machine. This means we don't have to worry about the specifics of the Turing machine; other models of computers give us no more computational power.

You don't really need to understand exactly what a Turing machine is to understand computation. Just think of any programming language that has access to an unlimited amount of memory. All programming languages are functionally equivalent, and all can compute exactly what can be computed by the simple Turing machine.

Turing back in 1936 showed that a Turing machine cannot in fact compute everything. The most famous example, the halting problem, says that no computer can look at some code of a program and determine whether that code will run forever or eventually halt.

During World War II, Alan Turing would play a major role in the code-breaking efforts in Britain. After the war he considered whether his Turing machine modeled the human brain. He developed what we now call the Turing test to determine whether a machine could exhibit human intelligence. Suppose you chat with someone through instant messaging. Can you tell whether the person you are chatting with is really a person or just a computer program? When a program can fool most humans, it will have passed the Turing test.

Unfortunately, Turing's research career was cut short. Turing was convicted under the then British law for acts of homosexuality in 1952. This ultimately led to Turing's suicide in 1954. It wasn't until 2009 that a British prime minister would make an official apology for Turing's conviction.

The Association of Computing Machinery named its highest award, the computer science equivalent of the Nobel Prize, after Turing for his work on the foundations of computer science and artificial intelligence. Many of the scientists mentioned in this chapter are Turing Award recipients.

Computational Complexity

In the 1950s digital computers started to play a significant role in society, and we needed some method for measuring the computational difficulty of solving various problems. The first was built on how scientists tried to describe the way humans think and communicate.

In 1943 two psychiatrists, Warren McCulloch and Walter Pitts, created a model called neural nets to describe brain activity. In the 1950s the logician Stephen Kleene created a limited version of the Turing machine called a finite automaton and studied the properties of the problems those machines could solve. Finite automata are very useful for specifying the control for simple machines (like a soda machine) but too limited to capture more complex algorithms.

In the 1950s the linguist Noam Chomsky wanted to understand how humans generated sentences in English and other languages. He developed methods, such as a "context-free grammar," that would create parse trees of sentences like that in the diagram below.

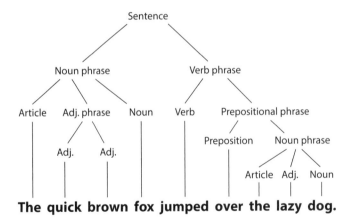

Figure 5-2. Parse Tree.

Context-free grammars do a reasonable but incomplete job of describing sentences in human language. The right kind of grammar to describe language—or whether one exists at all—remains a debate in linguistics.

Context-free grammars do turn out to be a great way to describe and parse programming languages and computer-generated data files such as XML. However, like finite automata, they fail to capture our intuitive notions of efficient computation.

Many other models were developed, but the big breakthrough came in 1962 from Juris Hartmanis and Richard Stearns working at the GE Research Labs in Schenectady, New York. They came up with a beautiful idea for understanding how well computer programs solve problems: see how the amount of time and memory needed by a computer algorithm changes as the problem description gets larger. Their 1965 paper "On the Computational Complexity of Algorithms" gave birth to the new field of computational complexity. Hartmanis and Stearns won the Turing Award for this work in 1993.

In the 1960s research in computational complexity ran in two different directions. One direction looked at what problems could and could not be solved in a given amount of time and with a specific amount of memory on very explicit models of computation. The other direction, spearheaded by Manuel Blum's doctoral research at MIT, took a very abstract approach, proving results that didn't depend on any specific model or even the specific type of resource such as time or memory. Neither of these approaches managed to capture the true essence of efficient computation.

P and NP

The right definition for efficient computation came in two papers in the mid-1960s, "Paths, Trees and Flowers" by Jack Edmonds and "The Intrinsic Computational Difficulty of Functions" by Alan Cobham.

Edmonds's paper will always be best known for giving the first efficient algorithm for the matching problem discussed in chapter 3. In a section titled "Digression," Edmonds also talks about the difference between exponential and algebraic order, though he cautions against any rigid criteria for efficiency:

> An explanation is due on the use of the words "efficient algorithm." . . . I am not prepared to set up the machinery necessary to give it formal meaning, nor is the present context appropriate for doing this. . . . For practical purposes the difference between algebraic and exponential order is more crucial

than the difference between [computable and not computable]. . . . It would be unfortunate for any rigid criterion to inhibit the practical development of algorithms which are either not known or known not to conform nicely to the criterion. . . . However, if only to motivate the search for good, practical algorithms, it is important to realize that it is mathematically sensible even to question their existence.

Edmonds's algebraic order is exactly P, the problems we can compute efficiently. As Edmonds suggests, we need a formal notion to define problems like whether P = NP, but we should keep in mind a more informal notion of efficient computation, a point of view I try to follow in this book.

Cobham independently defined P and talks about why the notion is a good one.

For several reasons, the class P seems a natural one to consider. For one thing, if we formalize the definition relative to various general classes of computing machines we seem always to end up with the same well-defined class of functions. Thus we can give a mathematical characterization of P having some confidence it characterizes correctly our informally defined class.

As with the definition of computability, the class P does not depend on the specific details of the computational model.

Cobham also offers a caution:

The problem is reminiscent of, and obviously closely related to, that of the formalization of the notion of effectiveness. But the emphasis is different in that the physical aspects of the computation process are here of predominant concern.

Perhaps Cobham realized there might be future models of computation that may not correspond to his class P. Later developments in randomized and quantum computation will show that perhaps we cannot have a fixed notion of efficient computation.

Then in 1971 came Steve Cook's paper that defined NP, the problems we can verify efficiently, the P versus NP problem, and the first NP-complete problem. A year later Richard Karp gave his paper showing that a number of important problems were NP-complete.

In 1972 the IBM T. J. Watson Research Center hosted the Symposium on the Complexity of Computer Computations, a meeting mostly remembered for Karp's presentation of his paper. The organizers of the meeting held a panel discussion at the end of the symposium on the future of the field. One of the questions asked was, "How is the theory developing from originally being a scattering of a few results on lower bounds and some algorithms into a more unified theory?" The panelists, including Karp himself, were at best only vaguely aware that the answer lay in Cook's and Karp's papers with the concepts of P, NP, reducibility, and what would later be called NP-completeness.

Karp realized they needed a good name for this field:

> "Computational Complexity" is too broad in view of the work of Blum and others, at least until we can gather them into our fold; "Concrete Computational Complexity" is too much like civil engineering; "Complexity of Computer Computations" doesn't ring true.

The name of the field would become *computational complexity*. The importance of the P versus NP problem quickly became apparent and greatly overshadowed most of the other research directions in the area. Abstract complexity felt less relevant. Manuel Blum himself changed directions and started working in cryptography and program checking. Blum received the Turing Award in 1995 for his broad range of research during the 1960s, 1970s, and 1980s. Asked years later about his change in research direction in the 1970s, Blum simply said, "Cook got it right."

The East

While the Russian "theoretical cybernetics" community had many important participants, we'll focus our story on three players, representing three ways of addressing perebor:

1. Sergey Yablonsky, who first focused on perebor for finding the smallest circuits for computing functions but whose power and arrogance worked against the development of computational complexity in Russia.

2. Andrey Kolmogorov, the great Russian mathematician, who measured complexity via algorithmic information.
3. Kolmogorov's student Leonid Levin, who independently developed the P versus NP problem and the theory of NP-completeness but for political reasons couldn't get a PhD in his native land.

Sergey Yablonsky

Research in computation in Russia went under the name theoretical cybernetics and didn't really get started until digital computers became an important part of the military in the 1950s. Sergey Vsevolodovich Yablonsky, born in 1924 in Moscow, went from serving in the Soviet Army in World War II to studying mathematics at Moscow State University. He received his PhD in 1953 under the supervision of Pyotr Novikov, who was one of the first Russians to study computability. Yablonsky teamed up with fellow Novikov student Alexey Liapunov to lead a series of seminars at Moscow State on the representations of logical functions. Their group became the center of research in computation theory in Russia.

The satisfiability problem that started off chapter 4 asks about logical formulas combining the basic operations of AND, OR, and NOT. We can actually express computation as a series of AND, OR, and NOT operations. Some problems need only a small "circuit" of ANDs, ORs, and NOTs and others require huge circuits to compute. Sergey Yablonsky in the early 1950s investigated this notion of what we now call circuit complexity.

Claude Shannon, an American who founded the field of information theory, showed that some logical functions require very high circuit complexity. Yablonsky looked at the difficulty of generating such functions. This sounds like a difficult task, but actually a strong version of $P \neq NP$ would mean that some simple-to-state search problems do not have small circuits that compute them.

Yablonsky first noted that by Shannon's results, randomly generated functions have near maximal circuit complexity. He then looked at the problem of finding such functions without using a randomized process.

Did it necessitate perebor, a brute force search through the space of all functions? He showed that any process that generated functions of near maximal complexity must embed ALL functions. Thus in some sense, any process that generates a hard function can be modified to generate any other function. Yablonsky argued that this means perebor was necessary since every function was generated when producing a hard function. Yablonsky titled his 1959 paper "On the Impossibility of Eliminating Perebor in Solving Some Problems of Circuit Theory."

Yablonsky did prove an important result but had a misleading interpretation. Just because you can generate any function from a hard function doesn't mean you have to generate all functions to get the hard functions. Yablonsky's result had little really to say about the computational complexity of finding hard functions. Yabonsky's student Zhuravlev in 1960 published a paper with an equally impressive title, "On the Impossibility of Constructing Minimal Disjunctive Normal Forms for Boolean Functions by Algorithms of a Certain Class," which also says little about algorithmic complexity. None of this research really applied to questions related to P versus NP.

Russia was a communist country with centralized control of the economy, including the academic profession. Yablonsky considered the perebor problem settled based on his own research and strongly discouraged continued research in this direction, particularly in the study of the role of computational complexity and algorithms. Yablonsky would become an influential member of important mathematical committees in Russia that dealt with coordination and control of mathematics, which would cause some controversy in the 1960s, as we'll see later.

Andrey Kolmogorov

Andrey Nikolaevich Kolmogorov was born in 1903 in the Russian city of Tamblov. In 1920 Kolmogorov studied at Moscow State University, but not initially mathematics. Instead, Kolmogorov tried his hand at history. He looked at the question of whether taxes in Russia in the Middle Ages were collected at the house level or at the village level. He analyzed tax data and showed that that data had a much simpler description if taxes were collected at the village level. He presented these results to the history faculty, to great applause. Asking them whether he could publish such a paper, he was told, "You only have one proof. You cannot publish in a history journal without at least two proofs of your hypothesis." And so Kolmogorov left history for a field where one proof suffices. Kolmogorov would go on to become the greatest Russian mathematician, if not the greatest mathematician anywhere, of the twentieth century, making fundamental contributions to nearly all areas of mathematics.

There is a possibly apocryphal story about how Kolmogorov saved probability theory in Stalin's Russia.

During the 1930s, Stalin, to consolidate his power, aimed to repress intellectual and artistic achievements that might challenge his regime. Stalin put pseudoscientists in powerful positions in the Russian biological community, and in 1937 large number of geneticists were arrested as "Trotskyite agents of international fascism." Genetics research, once a Russian source of pride, was effectively ended in the Soviet Union for several decades to come.

After attacking the geneticists, the lead "philosophers" in Russia went after the field of probability because of its notion of independent events. Probability theory measures the chances that some events will happen, such as the probability that two dice will sum to five is one out of nine. Probability gives a way of describing independent events. For example, when you roll two dice, the value of one die does not depend on the outcome of the other. This didn't fit well with Marxist philosophy, under which everything is interconnected and dependent on everything else.

These philosophers met with Kolmogorov and challenged him: "Under Marxism there cannot be independent events."

Kolmogorov realized the importance of his next words, for it could greatly affect the course of Russian mathematics. He responded, "But, you are mistaken." The agents believed they had Kolmogorov—to challenge the beliefs of determinacy meant an open challenge to the fundamental beliefs of Marxism, which could quickly lead to the end of Kolmogorov's career.

Kolmogorov went on, "Consider a priest that during a drought prays for rain. The following day it rains. These events are independent." The agents couldn't counter. Any dependence in these events would acknowledge the power of religion, which in itself would also be considered an attack on Marxist dogma. Kolmogorov had saved probability.

Kolmogorov's desire to understand probability and randomness led to one of his simple yet amazingly great ideas. Consider the following sequences of digits:

- 999 999 999 999 999 999 999 999 999 999 999 999 999 999 999 999 999
- 707 106 781 186 547 524 400 844 362 104 849 039 284 835 937 688 474
- 982 922 216 207 267 591 232 795 977 564 268 549 473 337 889 037 097

I chose one of these sequences via a random number generator and the other two from other means. Each of the sequences above has an equally likely chance of occurring, so why should any sequence be any less "random" than another? Try to guess which sequence was generated randomly before reading on.

A sequence of all 9's would seem to be a surprising output of a random number generator. The second sequence a few of you might recognize as the first fifty digits of the square root of one-half. Indeed it was the third sequence that was actually chosen at random.

Kolmogorov realized one could measure the amount of randomness of a sequence by the size of the shortest description of the sequence. The first sequence is just fifty 9's. The second sequence is $1/\sqrt{2}$. But the shortest way to express the third sequence is by just saying "982,922,216,207, 267,591,232,795,977,564,26854,947,333,788,9037,097." "Description" is an informal notion; Kolmogorov made it formal by expressing descriptions via computer programs.

These ideas were also developed independently by two Americans, slightly earlier by Ray Solomonoff (who despite the Russian name was born in Cleveland) and slightly later by Gregory Chaitin. Kolmogorov and his disciples took the concepts deeper, so this measure is often called "Kolmogorov complexity."

A random sequence is one whose shortest description is the sequence itself, for example,

982,922,216,207,267,591,232,795,977,564,268,549,473,337,889,037,097.

One can easily generate random sequences just by choosing numbers at random. Without random numbers these sequences are out of reach. No nonrandomized algorithm can produce arbitrarily long random sequences. I don't even know for sure whether

982,922,216,207,267,591,232,795,977,564,268,549,473,337,889,037,097

is truly a random sequence for I would have to test all shorter descriptions, many of which could be quite complicated.

Kolmogorov complexity has a rich and deep theory as well as several applications to machine learning, analysis of algorithms, and computational complexity. Kolmogorov complexity does not directly address the P versus NP problem, but thinking about Kolmogorov complexity led his student Leonid Levin directly to it.

Leonid Levin

In 1961, Liapunov left Moscow State University for Novosibirsk State University, where he established a department of theoretical cybernetics. Novosibirsk, located about 1,750 miles southeast of Moscow, is the third largest city in Russia and the largest in Siberia.

The department, staffed mostly by former students of Yablonsky and Liapunov from Moscow, quickly became a second major center of theoretical cybernetics in Russia. Boris Trakhtenbrot, already a senior researcher in the field at the age of forty, joined this center at its beginning and played a leading role. In 1962, Y. M. Barzdin received his PhD from the Latvian University in Riga and joined the Novosibirsk center. Trakhtenbrot started working with Barzdin on the basic concepts of computational complexity, attracting many students from both Latvia and Novosibirsk, and in the

1960s they started building an algorithmic theory of complexity similar to one being developed at the same time in the West.

But not all went well for research in computational complexity in Russia, as Trakhtenbrot wrote in 1984:

> We were upset by the deterioration in our relations with the "classical" cybernetics people, mainly Yablonsky. Their attitude to the introduction of the theory of algorithms into complexity affairs was quite negative. . . . They distrusted the role that computational complexity and algorithm complexity could play in the Perebor subject. These scientific divergences were likely intensified by the Perebor controversy, especially because at the time Yablonsky attained influential positions in the bodies that dealt with coordination and control of mathematical investigation.[*]

Kolmogorov visited Novosibirsk in the summer of 1963 and he and Trakhtenbrot shared their research, studying how the theory of algorithms could lead to an understanding of both information and complexity.

Soon after that visit, Kolmogorov went to Kiev University and visited a boarding school for high school–aged prodigies in math and physics. He threw out some problems, many of which were answered by then fifteen-year-old Leonid Levin. Kolmogorov later invited Levin to study with him at Moscow University. Levin as a graduate student produced two major lines of work.

Levin developed a universal search algorithm. Suppose Alice tells Bob she has an algorithm that solves clique problems quickly but won't tell Bob what that algorithm is. Levin's technique lets Bob create an algorithm that finds cliques almost as fast as Alice without Bob knowing Alice's algorithm. Levin developed his idea using a variation of Kolmogorov complexity that basically tries every algorithm in the right ways to achieve this result.

Thinking about search led Levin to think about problems that captured search, and he came up with the notion of "universal search problems," equivalent to the notion of NP-completeness developed by Cook. Levin devised a list of six universal search problems, including the satisfiability problem, and formulated the P versus NP problem.

---

[*] B. A. Trakhtenbort, "A Survey of Russian Approaches to Perebor Algorithms," *Annals of the History of Computing* 6, no. 4 (October 1984).

Levin had these two amazing results, the second of which is as strong as the paper that won Cook a Turing Award. Kolmogorov, though, didn't feel that either of Levin's results was publishable on its own, so he forced Levin to write them up together. The style of Russian mathematical publications at the time led to concise papers without details of the proofs of the results. Levin took this style to the extreme and wrote both results in a single two-page paper.

Levin applied for a PhD in Russia, though he didn't include that great work in his thesis. All young Russians were members of the Komsomol, the youth division of the Communist Party that controlled the Soviet Union. Levin was a troublemaker who often sabotaged some of the Komsomol activities. Levin misjudged the reaction to these activities and was denied his PhD officially for political reasons.

Levin could not sort out his political problems in Russia and managed to immigrate to the United States in 1978. Albert Meyer took him in as a graduate student at MIT, where Levin received a PhD one year later based on the great work he already had done. Levin then became a professor at Boston University, where he remains today.

By the time news of Levin's work reached the United States in the mid-1970s, the P versus NP problem had already drawn wide attention. Levin did not share in Cook's 1982 Turing Award, and it wasn't until the late 1980s that people started consistently calling the first results on NP-completeness the Cook-Levin theorem.

Relations between the Soviet Union and the United States started to open up in the 1980s. Alexander Razborov, a Russian student, played a major role in the development of circuit complexity as an approach to proving P ≠ NP, a story we tell in chapter 7.

After the collapse of the Soviet Union and the rise of the Internet, Russian mathematical researchers no longer worked in isolation. The world is now a truly global research environment.

The Gödel Letter

In 1956 Kurt Gödel wrote a letter to John von Neumann, one of the pioneers of computer science and many other fields. In this letter (written in German), Gödel talks about the satisfiability problem and formulates the

P versus NP question in different terminology. He suggests that if we lived in a world where P = NP, "the mental work of a mathematician concerning Yes-or-No questions could be completely replaced by a machine. . . . Now it seems to me, however, to be completely within the realm of possibility." This letter predates the papers of Cook and Levin by fifteen years.

We do not know if von Neumann responded or even saw this letter. Von Neumann had cancer at the time of the letter and passed away in 1957. The letter itself was unknown to the general scientific community until the late 1980s, well after the P versus NP problem was firmly established as a major open question. Gödel passed away in 1978 but was not in the right mental state in those last years of his life to notice that Cook had restated his question.

Since Gödel discovered the P versus NP problem years before Cook and Levin, why don't we call whether P = NP "Gödel's problem," and give him the credit? Science, for better or for worse, uses the Columbus principle: Christopher Columbus is famous not for being the *first* person to discover America but for being the *last* person to discover America. Gödel himself was not blameless. He didn't realize the importance of the question he raised with von Neumann and never mentioned it in a public venue. Cook and Levin still have the first published papers that state the P versus NP problem.

In honor of Gödel's fundamental work in logic and the Gödel letter, the theoretical computer science community initiated the Gödel Prize in 1993 to recognize recent outstanding research papers in the field.

The Martian Rule

How can we tell whether a scientific notion is natural, as God himself would have created it, as opposed to a product of human activity? Suppose we find a Martian civilization at about the same level of scientific progress that we have. If that civilization has a concept that is the same or similar to one of ours, then that concept should be natural, having been developed through multiple independent sources.

Of course, we don't have a Martian civilization to compare ourselves with, so we have to use our imagination. The Martians would have an

Exigius machine of computation that would be different from the Turing machine but have the same computational ability. The Martians would have their own version of the Church-Turing thesis that everything computable is computable by an Exigius machine. So the Turing machine itself is not natural, but computation is.

For P versus NP we have some non-Martian evidence. Researchers in Russia and North America with limited communication ended up developing the same P versus NP question and NP-complete problems. They had different motivation, in the East to understand the necessity of perebor and in the West to understand the power of efficient computation. Both ended up in the same place as did Kurt Gödel fifteen years before them.

Similarly, the hypothetical Martians would have developed the P versus NP question (or something very similar, assuming they haven't already solved it), marking the question as natural and important.

Chapter 6

— — — — — — — — —

DEALING WITH HARDNESS

IN CHAPTER 2 WE SAW THE BEAUTIFUL WORLD where P = NP, and life was good. We could optimize everything and learn anything. We had machines that could do just about anything we could imagine them to. Very beautiful, perhaps a little scary, and almost certainly a fantasy.

More likely we live in a dirtier world, the world where P ≠ NP, an "inelegant universe." Even if P = NP, until we find the algorithm that solves NP problems we might as well be living in the world of P ≠ NP. What about all those NP problems that we cannot solve efficiently? Do we just give up?

Sometimes we just have to solve hard problems. Harry works as a master scheduler at the Acme manufacturing shop. Harry's boss, Amy, tells him to schedule the machines to put together their latest cell phone, the A-Phone, and tells Harry to use as little time as possible. Harry has read the earlier chapters of this book and replies, "I'm sorry, but that problem is NP-complete. Many famous computer scientists don't believe there is any way to solve that problem quickly, so it is useless for me to try. I'm going bowling instead." Amy tells Harry to bowl well since he no longer has a job at Acme.

Amy quickly promotes George to Harry's position. Luckily for George, he has read this chapter of the book and creates a schedule for producing the A-Phone. Does George create a beautiful algorithm that will always find the best possible schedule? No. But does George get the job done? Yes.

NP-complete problems are nasty creatures, and if P ≠ NP we won't find fast algorithms that always produce the best solution that works for all possible problems. We don't have to give up, though, and this chapter will look at several approaches for dealing with hard problems.

For moderate-sized problems we can search over all possible solutions with the very fast computers we have today. We can use algorithms that might not work for every problem but do work for many of the problems we care about. Other algorithms may not find the best possible solution but still a solution that's good enough.

Other times you just can't get a solution for an NP-complete problem. You have to try to solve a different problem or just give up. There are other fish to fry.

Brute Force

Computers, even the ones that sit on your desk or in your pocket, are fast. Very fast. Incredibly fast. So fast that computers can often perform a "brute force" search through all the possible solutions of moderately sized problems.

It wasn't always that way. In 1971, the year that Steve Cook gave the P versus NP problem to the world, Intel released its first microprocessor chip, the Intel 4004. The Intel 4004 was the first complete CPU on a chip and the first commercially available microprocessor. The Intel 4004 ran at a then speedy 92,000 operations a second.

Now take Cook's satisfiability problem on twenty variables and consider the very simple algorithm of trying all possible settings of the variables to TRUE and FALSE. If it takes 100 steps to check each setting, then we can tell whether a formula is satisfiable in about nineteen minutes. Not too ridiculously long to solve, but there is not much you can express in twenty variables.

Solving a twenty-five-variable problem will take ten hours, a thirty-variable problem more than thirteen days. A forty-variable problem started in 1971 would have finished in 2009.

These days Intel releases many different models of processors each year. Let's pick one from 2009, the Intel i7-870, running at 2.93 billion operations per second (or more than 30,000 Intel 4004s). At this speed the forty-variable problem would take about ten hours. In other words, if you wanted to solve a forty-variable problem back in 1971, it would have been faster to twiddle your thumbs for thirty-eight years and then use 2009 technology than to solve your problem using 1971 technology.

For some other NP problems we can do ever larger examples. The NP-complete traveling salesman problem tries to find the best way to visit a large number of cities using the shortest total distance. Using something called the cutting-plane method we can solve traveling salesman problems on 10,000 cities in a short amount of time. Here is a solution for 13,509 cities of population at least 500.

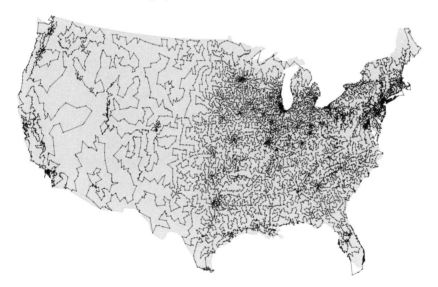

Figure 6-1. TSP 500.

For other NP problems, the number of possibilities is just too large, and one cannot hope to solve even moderate-sized problems.

While we are hitting physical limits on the speed of computers, we still expect the power of computers to continue to grow dramatically, with Moore's law reflected in the greatly increasing number of core processors on a chip. This will help us solve even larger NP-complete problems in the future. But the complexity of the problem also increases greatly with the problem size. Don't expect algorithms to solve all satisfiability problems on 150 variables or to find optimal traveling salesman tours on 20,000 cities any time in the near future.

Heuristics

Woodworkers in the seventeenth century would often use the width of their thumbs as a rough estimate of an "inch" in their measurements. Possibly from this practice we get the notion of a "rule of thumb," a simple procedure that gives us an imprecise but still quite usable method for determining the answer to some question. The famous saying "Red sun at night, sailor's delight; red sun in the morning, sailors take warning" does give a crude though often reliable way of predicting the weather. Moore's law itself has given us a rough method for predicting the power of future computers.

Computer algorithms are just another kind of procedure. A heuristic is an algorithm that works like a rule of thumb, not always giving the right answer but finding solutions for most of the problems you want to solve. People have created heuristics for NP-complete problems well before we knew they were NP-complete. Over the decades, we have developed a number of sophisticated heuristic procedures for a variety of hard problems. None of these procedures works on all instances for an NP-complete problem. But, depending on the context, many of these heuristic algorithms can solve the problems we need to solve.

Let us analyze in detail a simple yet powerful heuristic for map coloring. In chapter 3 we showed why the map of the United States requires four colors so that all bordering states have different colors.

California, Oregon, Idaho, Utah, and Arizona form a ring around Nevada. You need at least three colors to color California, Oregon, Idaho, Utah, and Arizona. Since they all border Nevada, that means you need a fourth color for Nevada. But Nevada is not the only such state. Kentucky is surrounded by Tennessee, Virginia, West Virginia, Ohio, Indiana, Illinois, and Missouri. Again, you need three colors to color Tennessee, Virginia, West Virginia, Ohio, Indiana, Illinois, and Missouri, and a fourth to color Kentucky.

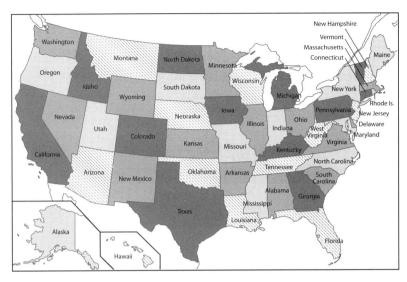

Figure 6-2. U.S. Map.

Take any map where there is some state surrounded by a ring of an odd number of other states and that map needs four colors.

Here is a map of the provinces of Armenia.

Figure 6-3. Armenia.

There are only two provinces that lie entirely within Armenia. Kotayk' has six neighbors and the capital province of Yerevan has four neighbors. Every state that doesn't border the ocean has an even number of neighbors. So the heuristic says we might be able to color this map with three colors, and we can.

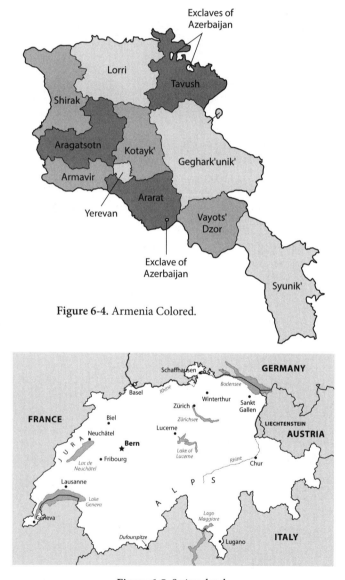

Figure 6-4. Armenia Colored.

Figure 6-5. Switzerland.

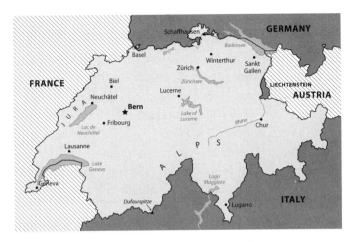

Figure 6-6. Switzerland Colored.

Switzerland has five neighbors, France, Italy, Austria, Lichtenstein, and Germany. Even though it has five neighbors we can still three-color this map.

Lichtenstein splits up the border between Switzerland and Austria, so, for the purposes of map coloring, we need to count Austria twice. It is not the number of neighbors but the number of borders that matters, and Switzerland has six, an even number of borders. Just the outside borders should count; we don't count any countries that lie wholly within another country (like the Vatican City in Italy).

If this heuristic always gives the right answer, then it gives us an efficient algorithm for map three-coloring, and since map three-coloring is NP-complete, we would have an efficient algorithm for all NP problems, and P = NP. So you should suspect at this point that the heuristic doesn't always work.

The heuristic will always work, with two exceptions.

1. The map has a lake that separates states, like Lake Michigan separating Illinois from Michigan.
2. The map has four or more states that meet up at a single point, like the four corners of Arizona, New Mexico, Colorado, and Utah.

In the United States the exceptions don't matter because we already know we need four colors to color the states, not even counting the separate blue color for the lakes.

The heuristic rarely fails in real-world examples. Let's instead look at the provinces of our made-up country, Frenemy. Every province borders an even number (four) of other provinces, yet it is impossible to color the provinces with only three colors (not including blue for the lakes) so that no two provinces colored the same share a border.

Each of these provinces borders exactly four other provinces. No province has a ring around it without a lake getting in the way. The heuristic says that one can color the eleven provinces with three colors, but in fact it is impossible to do so without coloring two adjacent provinces the same color. The heuristic fails to correctly tell us whether Frenemy is three-colorable.

The International Conference on Theory and Applications of Satisfiability Testing highlights research on all areas of the satisfiability problem, with a large emphasis on good heuristics. The conference organizes a SAT Race, a competition among different computer programs to see which do the best job on SATisfiability questions generated at random, questions specially crafted for their difficulty, and questions that arise from applications. Many of these algorithms can solve some satisfiability questions with over a million variables.

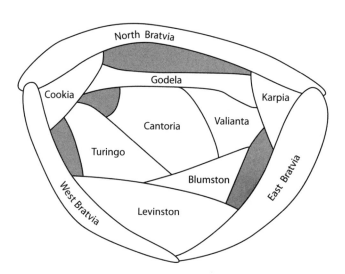

Figure 6-7. Country Frenemy.

Heuristics don't work all the time. No one gets close to a perfect score in the SAT Race. But often using various clever tricks, combined with lots of fire power from today's fast machines, can find solutions to even very large optimization problems.

Searching for Small Things

If we are looking for a clique of three people among the 20,000 inhabitants of Frenemy, we naively have to check a bit over a trillion possibilities, not too difficult for today's computers. But the numbers go up fast. To find a clique of four people, there are six quadrillion possibilities; a clique of five, 26 quintillion; and a clique of six, 88 sextillion (88 followed by twenty-one zeros). Pretty quickly these numbers get well beyond what our machines can handle.

For other NP-complete problems, searching for small solutions might be easier. A group in Frenemy is considered to be very cozy if for every two people who are friends in Frenemy, at least one of them is in the very cozy group.

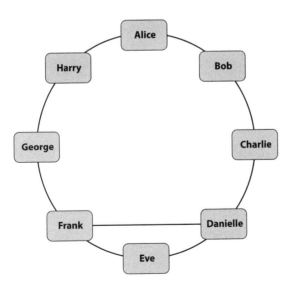

Figure 6-8. Very Cozy Group.

In this friendship diagram, Bob, Danielle, Frank, and Harry form a very cozy group because every friendship relations (all the lines in the diagram) have either Bob, Danielle, Frank, or Harry as one of the friends.

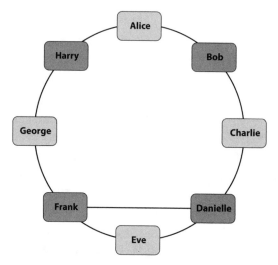

Figure 6-9. Very Cozy Group Four.

In this friendship diagram there are no very cozy groups of size three. For example, if we consider Alice, Charlie, and Frank, we miss the friendships between Eve and Danielle and between George and Harry.

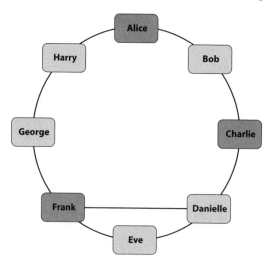

Figure 6-10. Very Cozy Group Three.

The very cozy problem, better known as "vertex cover," is one of Karp's original NP-complete problems.

Look at Frank in the friendship diagram above. If Frank was not in a very cozy group, then the group must contain George, Danielle, and Eve, since they all have friendships with Frank. If Frank had a hundred friends in Frenemy, either he would be a very cozy group or all of his friends must be in the very cozy group.

Using tricks like this, computer scientists can avoid looking at all possible groups when searching for small very cozy groups. Looking for a very cozy group of five people requires looking at about 100,000 possibilities. For ten people, 200,000 possibilities, for thirty people, 601,000 — all essentially taking no time on your laptop. You can search for a very cozy group of 113 people by searching less than a trillion possibilities in about the same time as searching for cliques with three people.

But wait! Didn't Karp show that finding the smallest very cozy group is NP-complete? It doesn't seem all that hard to find very cozy groups. Think about Frenemy and the vast friendship people have. It seems quite unlikely there is a very cozy group of only 113 people where for every pair of friends, one of them is in that group. The number of possibilities one has to search does start to get quite large as we allow the very cozy group to have a more reasonable size.

If you are looking for a group of 150 people, there are 1.5 quadrillion possibilities; for 200 people, over a sextillion; for 500 people, about 38 sexdecillion (38 followed by fifty-one zeros). So finding the smallest very cozy group in Frenemy is likely an impossible task. But if we just want to check that there are no very cozy groups of a hundred people, we can do that in a reasonable amount of time.

Approximation

Maybe we can't get the best possible solution to a problem, but often a not so bad solution is good enough.

Consider the traveling salesman problem, an NP-complete problem where one wants to visit a number of cities using the least total distance.

If I have to take a tour of fifty cities and the best tour of all them requires 1,800,000 miles and I can come up with a tour that needs 1,810,000 miles, I'm usually not going to pull my hair out over the 10,000 miles I could have saved.

On the other hand, if it costs me a dollar a mile to travel and I will get $1,805,000 in revenue for making the trip, the difference could mean going from $5,000 in profit to $5,000 in loss if my trip takes 1,810,000 miles instead of 1,800,000. If I can improve my trip again, maybe to 1,803,000 miles, I can be back in the black even though I never did find the absolute best route.

Even though the traveling salesman problem on a map is NP-complete and presumably difficult to solve exactly, we can find tours that get very close to the best solution. Sanjeev Arora and Joe Mitchell give an algorithm that subdivides the map into small pieces, finding good solutions for the traveling salesman problem in those small pieces and then connecting them all up again in a clever way.

Consider the map of 71,009 Chinese cities.

Figure 6-11. Chinese Cities.

We put a tight grid on top and solve the traveling salesman problem in each grid and piece them together. If there are too many cities in a small region, we will build smaller grids in those areas.

Using this scheme, we can find traveling salesman tours within a few percent of optimum in a reasonable amount of time.

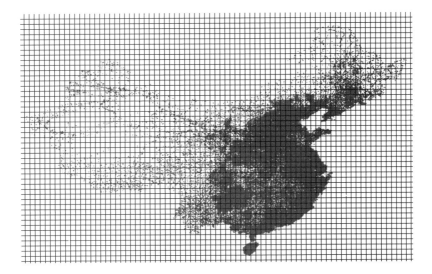

Figure 6-12. Chinese Cities Grid.

If all NP problems had such nice approximation schemes, the P versus NP problem would be nearly irrelevant. Life isn't that nice. Consider the clique problem, where we were trying to find a large group of people all of whom are friends of each other. We know no general procedures to get any significant approximation of the largest clique. Frenemy could have a clique of 2,000 people, and we might not be able to actually find a clique of size 15 in any reasonable amount of time.

If P = NP, we can find the absolute largest clique efficiently. This turns out to be the only way we can consistently find cliques even 0.1 percent as large as the largest clique. Any algorithm that can do better solving cliques approximately would make P = NP and allow us to solve cliques exactly.

Many NP problems are not as hard to approximate as clique or as easy to approximate as the traveling salesman problem on a map. Let's look again at very cozy groups.

In a small example like this we can just check all the possibilities and see the smallest possible very cozy group has four people, such as Frank, Danielle, Harry, and Bob. Suppose we didn't know that and let me describe a simple algorithm to approximate the smallest very cozy group.

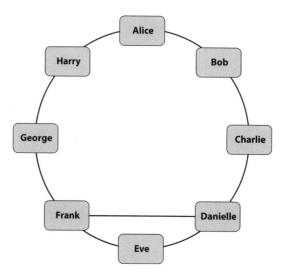

Figure 6-13. Very Cozy Group II.

First pick any pair of friends, say Alice and Harry, and mark them.

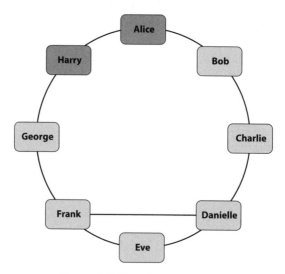

Figure 6-14. Very Cozy Group II 2.

Find another pair of friends, both of whom are not marked, and mark them as well.

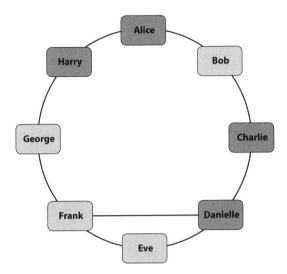

Figure 6-15. Very Cozy Group II 4.

Repeat until you cannot find two unmarked friends.

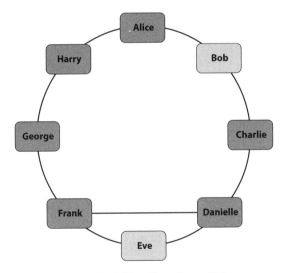

Figure 6-16. Very Cozy Group II 6.

At the end, the marked people will be a very cozy group.

In our example we have a very cozy group of size six. Every very cozy group must have at least one person from each of the chosen friendships. So every very cozy group has at least three people. So the best possible very cozy group has between three and six people.

This algorithm works on any friendship graph. If our algorithm found a very cozy group of size 100 (by adding 50 friendships), then we know the best possible very cozy group will have between 50 and 100 people.

We can always find a very cozy group at most twice the size of the smallest very cozy group. Can we do much better? Probably not.

If P = NP we can find the smallest possible very cozy group efficiently. But what if P ≠ NP? Then it would be impossible, in general, to find a very cozy group less than 36 percent larger than the smallest possible one. Any algorithm that can always find a very cozy group of size 36 percent larger than the smallest possible one could be converted to another algorithm that can solve any NP problem. This conversion process requires a series of very difficult results proved between 1990 and 2005.

The simple adding friendship algorithm we describe above finds a very cozy group at most twice, or 100 percent, larger than the size of the smallest possible group. If P ≠ NP, then the best we could hope for is about a 36 percent larger group. Do we have any hope of finding an algorithm that can find a group no more than 50 percent larger than the smallest very cozy group?

To answer this and other questions, computer scientist Subhash Khot developed a problem he called unique games, a variation on the map coloring problem where we have certain types of rules about how we can color neighboring countries. The unique games conjecture says the unique games problem is NP-complete. The jury is still out on whether or not the unique games conjecture is true.

If the unique games conjecture is true, we can limit even further our ability to approximate. Khot shows that if the unique games conjecture is true, one cannot find very cozy groups that are slightly less than twice as large as the smallest possible group. If the unique games conjecture is true, we cannot do significantly better at finding very cozy groups than by using the simple algorithm we describe above.

Solve a Different Problem

Sometimes no tricks, no matter how clever, can solve an NP problem. Then we try to solve a different problem.

Jane, a chef at Chez Turing in Palo Alto, needs a new tomato sauce to complement her famous spaghetti casserole. Chez Turing started the newest trend of nouveau restaurants, a "computational" school of food. Instead of experimenting directly with the food, Jane enters the color, taste, smells, and texture of the food she wants in a computer, which searches for the right combination of ingredients and how to put them together to get just the right point of yumminess while limiting cost and calories. So Jane asks for a deep red sauce at spice level 5, a texture like oatmeal but smoother, and a taste that tickles taste buds 5 and 11 with a smaller effect on the others that will perfectly complement the casserole.

The computer failed to find the right combination of ingredients that would achieve all of the requirements she gave to the system. So Jane called Tom, a computer genius who comes to aid Jane in exchange for an occasional free meal. Tom tried every heuristic he knew and a few he had to look up. He went to the cloud, renting computer time on Amazon's machines, to bring real raw computational power to the problem. When that fails he went to his friends in the Bay Area and offers a coveted reservation at Chez Turing to the first person who can find the ingredients to make the sauce. After about a week all give up—the problem is just too tough to solve.

Tom went to Jane to report the bad news. It was the first time he had failed to find the ingredients and process that Jane needed. Tom asked "Is there any other way, possibly a different taste that we could try to make the sauce?" Jane argued no, without this exact taste and texture the sauce would ruin the feeling of the casserole in the mouth. Tom asked "What about the casserole?"

Since finding the right sauce for the casserole was too hard a computational problem, Jane and Tom searched for a casserole and sauce that would work together. This task proved a bit easier and the computer spit out a recipe in only a few hours. By changing the problem, Chez Turing had a new hit dish on their hands.

In a different context, changing the problem is the bane of computer security specialists, who can use the latest cryptographic techniques to make things secure when the bad guys do only what's expected of them. But the bad guys succeed by doing the unexpected.

A smart card is like a credit card but with a little processor inside, as well as a secret key embedded in its memory. Smart cards can be used to verify identity or even store money that you can use at a store or, say, a parking meter to transfer money without having to make a connection to a centralized computer. Even if one observes the data going back and forth between the card, it can be quite difficult to figure out the secret key.

Suppose Thomas works at a hotel and gains access to Anne's smart card while she is swimming in the pool. Thomas has a specialized card reader that can try sending different data to the smart card and sees what data come back. If $P = NP$, then the thief can use this input/output behavior to find the key. If $P = NP$, these smart cards are quite useless. Since we likely live in a world where $P \neq NP$, finding the key could be difficult. Smart card protocols are specifically designed so that the techniques in this chapter, like brute force and heuristical algorithms, will make no headway in finding the secret key. Failing to find the key, Thomas normally would just return the card to Anne, or when she finds out the card is missing she will call her bank to cancel the card.

Thomas is no normal identity thief. He knows he can use the smart card in ways its designers never imagined. Thomas throws Anne's smart card into a microwave for a few seconds. Don't try this at home: a microwave will make a smart card or any other piece of electronics next to useless. Even a short microwave burst can fuse a few circuits. and if it doesn't permanently damage the smart card, it will cause the smart card to have some unexpected behavior.

Unexpected behavior is good for Thomas. Thomas has no idea how to find the key if the smart card works as it is supposed to. But a slightly damaged smart card still will give some data out when data is put in, just not the same kind of input/output behavior that the card had before.

The new input/output behavior still depends on the secret key but no longer in a way designed to make finding the key difficult. Thomas can now use heuristical and brute force approaches with some hope that they might work, and find the secret key.

Once Thomas finds the key, he can use it to make two new cards, each of which is identical to Anne's pre-microwaved card. Thomas returns one of these cards to Anne and keeps one for himself. With careful use, Thomas can steal money from Anne's account, and it might be weeks or months before Anne and her bank discover the problem.

Thomas found the secret code by changing the problem from one NP problem specifically designed to be difficult to another problem not designed at all. Newer smart cards have mechanisms to prevent the microwave attack, but we can never underestimate the imagination of people intent on finding secret information.

Acceptance

Sometimes a problem just defies solution, and you have to give up and do something else.

Richard has stumbled on the right chemical formula to create a super-drug that will allow him to take over the world, or at least the greater Cincinnati area. He works out a plan to infect the Miller water treatment plant with his formula, which will render the inhabitants very relaxed and easy to manipulate so that Richard can quickly step in and control their society.

All Richard has to do is collect enough of the appropriate chemicals to make his dream a reality. Unfortunately, an earlier incident led to an injunction that prevents Richard from ordering the chemicals directly from a supply company. So Richard starts analyzing various household products that he can break down and use for his concoction. Richard inputs the various parameters into his laptop: the cost of the products, the limits he can safely buy at each of the twenty Walmarts and fourteen Targets in the greater Cincinnati area, the cost of transportation and storage locations, of which he will need several. There were many different combinations of products he could use. He puts these parameters into the computer, but there are just too many possibilities to check to find one that will fit his time and monetary restrictions. He knows in his heart that there is some way he could make it all work, but try as he might, using every heuristic and approximation algorithm that he can find on the Internet, nothing leads to a scheme that will get Richard the chemicals he needs. Eventually Richard gives up and returns to his job as a security guard at the local toothpaste factory.

So the difficult complexity of NP problems saves Cincinnati.

Putting It Together

Typically, no single technique will suffice to handle the difficult NP problems one needs to solve. We often have to try several of the techniques described in this chapter. We can't solve the problem we want to solve, so we try to solve a different problem. But even that different problem may still be NP-complete, so we try some heuristics to attack it. The heuristics rarely solve the problem, but they may give good enough approximate solutions.

If P = NP, all these issues go away, and we can do great things with a single algorithm. But if, as we suspect, P ≠ NP, we can almost always still get something. It might take significantly more effort, we might not solve quite the problem we care about, and we might not get the very best solutions, but if it still gets the job done, that's good enough.

Chapter 7

— — — — — — — — — — —

PROVING P ≠ NP

JURIS HARTMANIS, ONE OF THE FOUNDERS of computational complexity, has a saying: "We all know that P is different from NP, we just don't know how to prove it."

In the earlier chapters we explored the P versus NP problem, what it is and why it matters, the beautiful but unlikely world where P = NP, and how to deal with hard problems if P and NP are different.

The P versus NP problem is also an amazingly challenging mathematical question. Almost immediately after Cook, Karp, and Levin brought the problem and its importance to the world, computer scientists and mathematicians tried to formally prove that P = NP or P ≠ NP. All the usual techniques failed to work. By the end of the 1970s the general

Figure 7-1. DILBERT © 1997 Scott Adams. Used by permission of UNIVERSAL UCLICK. All rights reserved.

consensus was that "substantially new proof techniques will probably be required to resolve the P versus NP problem."

The following decades saw incredible advances in both computer science and mathematics, including the resolution of the most famous of all open mathematical questions, Fermat's Last Theorem. Around 1637 Pierre de Fermat, a French lawyer and amateur mathematician, wrote the following in the margins of his copy of the *Arithmetica* (an ancient Greek textbook):

> I have discovered a truly marvelous proof that it is impossible to separate a cube into two cubes, or a fourth power into two fourth powers, or in general, any power higher than the second into two like powers. This margin is too narrow to contain it.

In other words, there are no natural numbers a, b, and c all greater than zero and n greater than 2 such that $a^n + b^n = c^n$. Fermat never mentioned this proof again, so it is likely he never had a true solution. The problem gained great notoriety as it became the classic unsolvable math puzzle. Kids like me dreamed of being the first person to solve this famous problem. One of those kids grew up and did just that. In 1994 the Princeton mathematician Andrew Wiles, building on a long series of papers in number theory, developed a proof of Fermat's claim and became an instant celebrity, at least as much of a celebrity as a mathematician could be.

This chapter won't show how to solve the P versus NP problem, or this would have been a very different book. Instead we explore a few of the ideas that people have tried to resolve the P versus NP problem. Alas, these ideas have not panned out to anything close to a solution to the problem. To prove P $\neq$ NP one needs to show that no algorithm, even those that haven't been discovered yet, can solve some NP problem. It is simply very difficult to show that something can't be done. But it is not a logically impossible task. So we keep hoping that someday we will see a solution to perhaps the most important and challenging of all mathematical questions.

The Liar's Paradox

Consider this puzzling sentence.

This sentence is not true.

Figure 7-2. Sentence Box.

Is the sentence true or false? If the sentence were false then the sentence cannot be not true, a double negative that means the sentence is true. If the sentence is true that would mean the sentence is not true but false. Either way we get the opposite of our assumption. This paradox is often called the liar's paradox. I am lying right now. Or was I?

There are no true paradoxes in mathematics, just bad mathematics. One cannot mathematically formulate the statement "This sentence is not true" because a sentence cannot talk about its own veracity.

Around 1930, Kurt Gödel discovered that mathematics can talk about proofs and that he could create mathematical statements that stated whether other statements had a proof of correctness. Gödel discovered how to have a sentence talk about its own proof and created a mathematical formula that expresses a variation of the above.

There is no proof that this sentence is true.

Figure 7-3. Proof Box.

Suppose the sentence is false. Then there is a proof that the sentence is true. This contradicts our assumption that the sentence is false. So the sentence must be true.

Have we arrived at another paradox? Not quite. The sentence is true; however, there is no proof that the sentence is true. In one fell swoop, Gödel upended the foundations of mathematics: there are true statements that one cannot prove are true.[*]

Suppose someone tells you he has a magic box that can predict the future. Ask him to tell whether you will punch him with your right hand or the left. If the box says left hand, then hit him with your right. If the box says right, then give him the left hook. Either way the box was wrong.

---

[*] Didn't we just prove the sentence is true? Not quite: we had to assume that everything we can prove true is true. So Gödel also shows that we cannot prove that "everything we can prove true is true" unless we can prove false things. That's what you get by reading footnotes.

The same can be done for computation. We have all had the experience of a computer screen that just shows the hourglass, and we don't know whether the computer is stuck or whether it is just taking a long time to think. Should you reboot the machine or should you wait longer? Wouldn't it be nice if someone came up with an algorithm that could tell us if the computer was stuck in some kind of infinite loop! It would be nice, but it's also impossible.

To understand why, we begin with a simple observation first realized by Alan Turing in his classic 1936 paper that started the field of computer science: a computer program is just data, just like your documents and spreadsheets. One program can analyze the code of another program.

A computer program either will eventually finish its computation and output some answer or it will compute forever. Suppose we have an algorithm that can tell whether a program will eventually finish. We then run that program on itself.

Either the program inside the box finishes or it doesn't. Either way we

---

If the program inside this box finishes then run forever.

Figure 7-4. Program Box.

get a contradiction. So our assumption is wrong, and we cannot have a program that can tell whether another program finishes. Not on a PC, not on a Mac. Not now, not in a hundred years. There simply can never be any program that can tell whether another program finishes.

Can we use a similar idea to show one cannot solve some problems efficiently, that is, show that the problem does not lie in the class P, the class of efficiently solvable problems? Indeed we can.

We start by creating an algorithm Q that works as follows, using data as the code for program R.

- If program R, using as data the code for R, outputs "Yes" in an efficient way, then Q outputs "No."
- Otherwise Q will output "Yes."

Suppose we have some efficient algorithm S. Then Q(S) says "Yes" exactly when S(S) says "No." So no efficient algorithm can have exactly the same behavior as Q.

Q is still a legitimate algorithm. So the problem that Q solves is computable but not efficiently computable.

If we could show Q is in NP, that is, has efficiently verifiable solutions, then Q would be in NP and not in P. This means P ≠ NP, solving that great question.

We don't know, and in general we don't even believe, there is an NP algorithm for Q. For this and other reasons, this paradox approach to P versus NP seems doomed to fail, at least as a direct attempt at showing P ≠ NP.

Circuits

At the heart of any modern computational device is the integrated circuit.

The integrated circuit consists of millions or billions of transistors, tiny devices that amplify and switch electrical impulses. These devices implement *gates*, simple operations on wires that can carry an electrical charge.

Let's start with the simple wire. Either the wire carries a high voltage or it carries a low voltage. A wire can have only one of two values,

Figure 7-5. Circuit. Courtesy of University of Pennsylvania, ESE Dept.

which are usually indicated by on and off, one and zero, or true and false. These two-valued systems are what we call a bit, short for "binary digit."

There's not much a wire can do by itself, or even with other wires. To create computation, we have to manipulate the information on the wires. The simplest way is just to flip the value of the wire, a NOT gate.

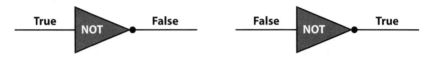

Figure 7-6. NOT.

If there is voltage on the left of the NOT gate, then there will be no voltage on the right, and vice versa.

The real power of computation comes not from manipulating a single wire but from combining wires in a useful way. An AND gate takes two or more wires and combines them into a single value that is true if both of the incoming wires are true. An OR gate takes two or more wires and combines them into a single value that is true if at least one of the incoming wires is true.

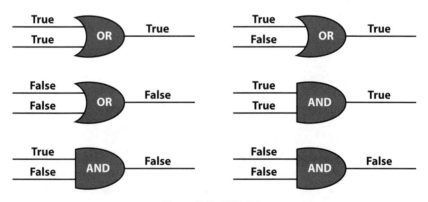

Figure 7-7. AND-OR.

We can use these simple building blocks to create more complicated operations. For example, the Exclusive-OR of two input wires is where the output wire is true if exactly one of the input wires is true.

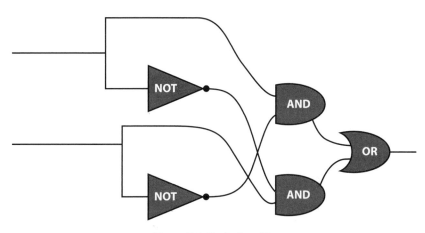

Figure 7-8. Exclusive-OR.

Every function, no matter how complicated, can be computed by a circuit of AND, OR, and NOT gates. Let's go back to the clique question in Frenemy, where we wanted to find a group of people who are all friends with each other. We can use an AND gate to tell if Alice, Bob, and Carol are all friends with each other.

Figure 7-9. 3 AND.

Suppose we want to know if there is a clique of three people among Alice, Bob, Carol, David, and Eli. We build a circuit for that in figure 7-10.

This circuit has ten AND gates representing the ten possible cliques of size three that can be found among five people. Suppose we are looking for fifty- person cliques among the 20,000 people of Frenemy. A straightforward circuit would have

3,481,788,616,673,927,174,126,704,329,430,033,822,428,741,136,969,
800,292,509,234,146,086,195,855,824,730,457,281,170,250,134,309,383,
506,694,008,809,661,825,431,661,561,845,957,650,386,210,936,569,600

AND gates, an impossibly large circuit.

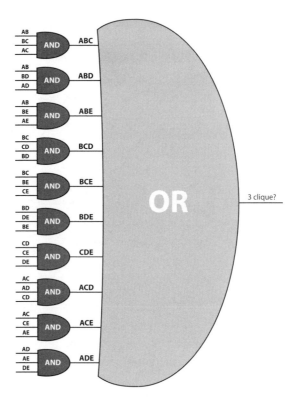

Figure 7-10. Clique Circuit.

What does this have to do with the P versus NP problem? Every problem in P—that is, every problem that has a computationally efficient solution—also has a reasonably small circuit using AND, OR, and NOT gates that can solve the problem. If we can show that some problem in NP, like clique, cannot have small circuits, then NP ≠ P.

Can one prove that clique does not have small circuits? This is closely related to the P versus NP problem and just as unknown, though most computer scientists believe clique does not have small circuits, just as they believe P ≠ NP.

Look at the circuits we created for clique above. Notice they have no NOT gates. Not all problems can have circuits without NOT gates; for example, it is impossible to do a simple Exclusive-OR without using a NOT gate. The clique problem, on the other hand, can be computed with circuits just using AND and OR gates, albeit very big ones.

In 1985 Alexander Razborov, then a student at the Steklov Mathematical Institute in Moscow, showed that any circuit that solves the clique problem must have a large number of gates if the circuit only has AND and OR gates (and no NOTs). This didn't solve the P versus NP question. You don't need NOT gates for a circuit to compute the clique problem, but possibly using NOTs could greatly reduce the number of gates you need for a clique circuit.

Nevertheless, Razborov's result was viewed as a huge breakthrough toward solving the P versus NP problem. My PhD adviser, Michael Sipser, said at the time that a solution to the P versus NP problem was just around the corner. All you had to do was modify Razborov's techniques so they could handle NOT gates as well. Then clique, an NP problem, would require large circuits using the full range of AND, OR, and NOT gates. Clique could then not have an efficient algorithm since all efficient algorithms can be converted to small circuits. So clique would be a problem in NP but not in P, proving P ≠ NP. Unfortunately, life was not that simple.

Razborov's papers arrived in the United States in the original Russian. Sipser gathered together some Russian students and they translated the papers carefully, hoping beyond hope that the next paper would be that one that finally put the P versus NP problem to rest. Razborov produced many other nice papers, but showing P ≠ NP this way was not to be.

In chapter 3 we described the matchmaking problem in Frenemy, where we wanted to match up the residents of Frenemy into pairs of friends. Matchmaking, like clique, has circuits that use only AND and OR gates. The techniques used to show that clique requires a large number of AND and OR gates also works for matchmaking. Any circuit consisting of only AND and OR gates that solves the matchmaking problem requires a large number of gates.

Unlike with clique, we know very efficient algorithms to solve the matchmaking problem. So there are small circuits that solve matchmaking using AND, OR, and NOT gates. The NOT gates are not needed to solve matchmaking, but using them dramatically decreases the number of gates needed to compute solutions to the matchmaking problem. The lowly and simple NOT gate is much more powerful than it appears.

That put a big crimp in the attempts to use Razborov's result on cliques to settle the P versus NP problem. Showing that a problem required large circuits with only AND and OR gates did not mean that the problem still required large circuits when we added NOT gates. Some of Razborov's later work clarified these issues. He showed exactly why his proof falls apart and is most likely unfixable in the presence of NOT gates.

Later, Razborov and Steven Rudich of Carnegie-Mellon developed a notion of "natural proofs." Natural proofs characterized a wide range of proof techniques for circuits and gave some strong evidence that these techniques would never solve the P versus NP problem.

We have seen some small progress with circuits using non-"natural" techniques building on the paradox ideas talked about earlier in this chapter. But hope for a solution to P versus NP by showing an NP problem requires large circuits has dimmed considerably.

How Not to Prove P ≠ NP

On August 6, 2010, Vinay Deolalikar, a research scientist at Hewlett-Packard Labs, sent to twenty-two leading theoretical computer scientists a copy of his paper, titled simply "P ≠ NP." Many people, dreaming of fame and fortune (the million-dollar bounty from the Clay Mathematics Institute), have come up with "proofs" that settle the P versus NP problem, claiming to show P ≠ NP, or P = NP, or that it is impossible to determine whether or not P = NP, or that the P versus NP question doesn't even make sense. Every year dozens of these manuscripts are emailed to computer scientists, submitted to academic journals, and posted on the Internet. The most prestigious computer science journal receives a steady stream of papers claiming to resolve the P versus NP question and has a specific policy for those papers:

> The Journal of the ACM frequently receives submissions purporting to solve a long-standing open problem in complexity theory, such as the P/NP problem. Such submissions tax the voluntary editorial and peer-reviewing resources used by JACM, by requiring the review process to identify the errors in them. JACM remains open to the possibility of eventual resolution of P/NP and related questions, and continues to welcome submissions on the subject. However, to mitigate the burden of repeated resubmissions of

incremental corrections of errors identified during editorial review, JACM has adopted the following policy: No author may submit more than one paper on the P/NP or related long-standing questions in complexity theory in any 24 month period, except by invitation of the Editor-in-Chief. This applies to resubmissions of previously rejected manuscripts.

Most attempts at solving P versus NP are unreadable or clearly wrong and are generally ignored by the community. Deolalikar's paper went through a different process. Deolalikar had published research papers in the past, and his paper was much better written than most P versus NP manuscripts. Some theorists felt that his paper deserved a closer look. This led to an Internet frenzy of tweets and blogs that prematurely announced the P versus NP problem had been solved. A few famous computer scientists and mathematicians took a close look at Deolalikar's paper. They found several flaws, minor, major, and fatal. By August 16, a mere ten days after Deolalikar circulated his paper, the *New York Times* had published an article, "Step 1: Post Elusive Proof. Step 2: Watch Fireworks," describing the whole affair. By this time the general consensus was that the proof was wrong and could not be fixed. The status of the P versus NP problem remains open.

Hopefully, this book has inspired many of you to understand the importance of the P versus NP problem, and you may yourself be tempted to try to solve the problem. I encourage you to try, for you cannot truly understand the difficulty of a problem without attempting to solve it. Keep in mind that this book does not formally define the P versus NP problem, and if you really want to tackle it you will need to understand the precise description of the problem. The website for this book gives several good sources for a technical formulation of the P versus NP problem and the attempts to settle this question.

Suppose you have actually found a solution to the P versus NP problem. How do you get your $1 million check from the Clay Mathematics Institute? Slow down. You almost surely don't have a proof. Realize why your proof doesn't work, and you will have obtained enlightenment.

Let me mention a few of the common mistakes people make when thinking they have a proof.

Perhaps the first bad P ≠ NP proof goes back to 1550 and the writings of Gerolamo Cardano, an Italian mathematician considered one of the founders of the field of probability. Cardano, in creating a new cryptographic system, argued for the security of his system because there were too many keys to check them all. But his system was easily broken. You don't have to check all the keys when doing cryptanalysis on secret messages.

Variations on Cardano's basic error recur in many modern attempts at proving P ≠ NP. Consider the clique problem again. Any algorithm that tries to solve the clique problem and can't find a solution must guarantee that no clique of a given size exists. The only way such an algorithm can work is by checking every possible set of people and seeing if they form a clique. Since there are very many sets, all algorithms must take a very long time, and so there are no efficient algorithms to solve clique. Thus P ≠ NP. Q.E.D. (Q.E.D. is short for the Latin phrase *quod erat demonstrandum,* or "what was to be demonstrated," and is often inserted at the end of an argument to signify a mathematical proof is completed.)

There are many variations of this type of reasoning. The faulty logic lies in "the only way." An algorithm can work in mysterious ways. It doesn't have to work along the lines you think it should or even respect the structure of the problem at all. Perhaps the algorithm converts the friendship diagram to some weird algebraic expression that has a certain kind of solution only if a clique exists. It seems unlikely, but you can't rule it out just because you don't think an algorithm can work that way.

It is very hard to argue against someone using this proof technique. They will often come back and ask me for an algorithm for clique that doesn't work that way. Of course, I can't provide one, because that would imply P = NP, which likely isn't even true. But the burden of proof (no pun intended) is on you, the person who claims he (or she, though the people who give bad proofs have been almost exclusively male) has a proof. You need to argue that no algorithm of any kind exists.

To prove P = NP, you need "just" to show some NP-complete problem has an efficient algorithm. So people will give an algorithm for something like the clique problem and say they have proved P = NP. But an algorithm alone does not constitute a proof. One must give a formal proof that on every *instance* of the problem, that is, all possible friendship diagrams, the algorithm both runs efficiently and gives the correct answer. All attempts at such algorithms either don't give a proof for all possible cases or the proof itself is incorrect or incomplete. These algorithms typically give wrong answers or run inefficiently on complicated instances.

The Present

We are further away from proving P ≠ NP than we ever were. Not literally, but in the sense that there is no longer any obvious path, no known line of reasoning that could lead to a proof in the near future.

The only known serious approach to the P versus NP problem today is due to Ketan Mulmuley from the University of Chicago. He has shown how solving some difficult problems in a mathematical field called algebraic geometry (considerably more complex than high school algebra and geometry) may lead to a proof that P ≠ NP. But resolving these algebraic geometry problems will require mathematical techniques far beyond what we have available today. A few years ago Mulmuley felt this program would take 100 years to play out. Now he says that is a highly optimistic point of view.

How long until we see an actual resolution of the P versus NP problem? Maybe the problem will have been resolved by the time you read this. Far more likely the problem will remain unsolved for a very long time, perhaps longer than the 357 years needed for Fermat's Last Theorem, or perhaps it will forever remain one of the true great mysteries of mathematics and science.

Chapter 8

— — — — — — — — —

SECRETS

WE ALL HAVE SECRETS, from passwords to emails we don't want the world to see. If P ≠ NP, then some NP problems have secrets, solutions that we can't find quickly. In 1976, Whitfield Diffie and Martin Hellman suggested that we could use NP to hide our own secrets. The field of cryptography, the study of secret messages, changed forever.

A Very Short History of Classical Cryptography

People have been sending secret messages as long as there have been messages to send. Julius Caesar used a simple substitution cipher in which each letter is replaced by the one three letters later.

"The early bird gets the worm" becomes "Wkh hduob elug jhwv wkh zrup." Methods of encrypting by rotating the letters became known as Caesar ciphers.

These techniques worked well in ancient Rome as the encoded messages looked just like random gibberish, and they had limited means

Old	A	B	C	D	E	F	G	H	I	J	K	L	M	N	O	P	Q	R	S	T	U	V	W	X	Y	Z
New	D	E	F	G	H	I	J	K	L	M	N	O	P	Q	R	S	T	U	V	W	X	Y	Z	A	B	C

Figure 8-1. Caesar Cipher.

to break the code. By the ninth century mathematicians had developed methods of looking at the frequencies of letters and short words to determine the original message. One could look at "Wkh hduob elug jhwv wkh zrup" and notice that the letter "h" occurs four times, more than any other letter. The letter "e" occurs the most often in the English language, about 12 percent of the time. So you could (correctly) deduce that the letter "h" encodes the letter "e." Then you might notice "wkh" occurs twice, and you know "h" should be "e," so you may deduce "wkh" is "the." You are well on your way to decoding the message.

In the fifteenth century, during the Italian Renaissance, Leon Battista Alberti developed a more complicated method known as polyalphabetic cipher, in which different substitutions were used in different parts of the messages. These codes remained mostly unbreakable until the nineteenth century, when more systematic attacks on codes were developed.

The author and poet Edgar Allan Poe had also developed a skill as a cryptanalyst. In 1839 he challenged the public to provide him with difficult-to-break ciphers, and Poe would provide the solutions. A year later Poe published an essay, "A Few Words on Secret Writing," in which he claimed "human ingenuity cannot concoct a cypher which human ingenuity cannot resolve." His 1843 story "The Gold-Bug" revolves around breaking a secret message, one of the first important works of fiction depicting secret codes.

In 1903 Arthur Conan Doyle published *The Adventure of the Dancing Man,* in which Sherlock Holmes breaks a substitution code based on stick figures.

With the mechanical age came various devices to create codes to encrypt and decrypt secret messages. Perhaps the most famous of these, the Enigma machine, was first developed by Arthur Scherbius in 1918 in Germany.

The Enigma machine has several rotors each of which scrambles letters as they are typed. The rotors step along at different rates so that a

Figure 8-2. Dancing Men.

Figure 8-3. Enigma Machine.

different substitution is done for each letter. This machine creates a very complex version of Alberti's polyalphabetic cipher that is very difficult to crack.

The Enigma was the primary source of codes for the Germans during World War II. Just before the war, Polish military intelligence provided the British with descriptions of the Enigma and some deciphering techniques. The British government set up a secret project, code-named "Ultra," to break these codes. Their recruits included people who excelled at crossword puzzles and chess, as well as great mathematicians. These operations involved Alan Turing, the father of computation, and led to the development of Colossus, the first programmable digital computer. According to Winston Churchill, "It was thanks to Ultra that we won the war."

Cryptography has been and will always remain a game of cat and mouse, between the code makers and the code breakers. But in the 1970s the game greatly changed as computer scientists started building codes based on the difficulty of solving some NP problems.

Modern Cryptography

"We stand today on the brink of a revolution in cryptography." That's the start of a famous 1976 paper by Whitfield Diffie and Martin Hellman from Stanford. "The development of cheap digital hardware has freed [cryptography] from the design limitation of mechanical computing and brought the cost of high-grade cryptographic devices down to where they can be used in such commercial applications as remote cash dispensers and computer terminals."

Diffie and Hellman realized that computers could make complex protocols cheaply available as software, but computers also brought new challenges. Computer-to-computer networks would become commonplace, and efficient and low-cost methods to secure communication over these networks would be needed. Talking about the development of the P versus NP problem earlier that decade, Diffie and Hellman stated, "At the same time, theoretical developments in information theory and computer science show promise of providing probably secure cryptosystems, changing this ancient art into a science."

Before Diffie and Hellman, to decrypt a secret message you needed the same key as the one used to encrypt the message. Both parties would have to meet in advance to agree on a secret key. If a general needed to send a message to a lieutenant in the field using Enigma machines, they both had to know the settings for that machine, particularly how the rotors were initially set at the beginning of the message. The lieutenant would carry a book of settings, usually one for each day. If this book fell into enemy hands, secret communications would come to a halt until a new set of books could be distributed. Soldiers knew to take great care of these books and destroy them if captured, so new books were not needed very often.

A computer network creates new challenges, as we can't assume that these networks are secure. Computer networks in the late twentieth century often traveled through normal phone lines, easy to tap into. Today someone sitting on the other side of a coffee shop can read all the data you send using the shop's Wi-Fi.

We can't send the key through the network or it would compromise future communication. You would have to somehow physically send a secret key to the other person before you could start sending secret messages. That could get very expensive and time-consuming.

Diffie and Hellman, building on earlier work of Roger Merkle, proposed a method to get around this problem using what they called "public-key" cryptography. A computer would generate two keys, a public key and a private key. The computer would store the private key, never putting that key on the network. The public key would be sent over the network broadcast to everyone.

Diffie and Hellman's idea was to develop a cryptosystem that used the public key for encrypting messages, turning the real message into a coded one. The public key would not be able to decrypt the message. Only the private key could decrypt the message.

If Diffie wanted to send a secure message, "Attack at noon," to Hellman, first Hellman would create private and public keys. Hellman would send the public key to Diffie (and to anyone else who happened to be listening in) and keep the private key to himself. Diffie would then use Hellman's public key to encrypt the message "Attack at noon" to "tzljcnnfekktis" and send the encoded message "tzljcnnfekktis" to Hellman. Diffie wouldn't need the private key for the encryption, just the public key. The eavesdropper would see the encoded message "tzljcnnfekktis," but even with the public key the eavesdropper would not be able to recover the original message. Hellman could use his private key on "tzljcnnfekktis" to recover the original message, "Attack at noon."

Is public-key cryptography even possible? Not if P = NP. If P = NP, there would be an efficient algorithm to recover the correct private key from the public key.

Even by 1976 most computer scientists believed that P ≠ NP, which suggested that public-key cryptography might be possible. Diffie and Hellman suggested such a system, but their protocol is not as commonly used as a scheme discovered by three computer scientists, Ronald

Rivest, Adi Shamir, and Leonard Adleman, in 1978 and named "RSA" after them.

RSA builds on the idea that it is very easy to multiply numbers but seemingly very difficult to factor them. If you pick two large prime numbers, say 5,754,853,343 and 2,860,486,313, it is easy to compute the product, 16,461,679,220,973,794,359. On the other hand, it seems difficult to invert this process, to take 16,461,679,220,973,794,359 and recover the factors 5,754,853,343 and 2,860,486,313. RSA uses much larger numbers, typically hundreds of digits long. We can't prove that factoring is computationally difficult without proving P ≠ NP, but most people believe factoring is a difficult problem.

Rivest, Shamir, and Adleman shared the 2002 Turing Award for their protocol.

In an odd twist, it turns out that the RSA protocol was first discovered in 1973 by Clifford Cocks of the Government Communications Headquarters, the British version of the National Security Agency. This fact remained classified until 1997.

You probably use RSA every day. Just look at a typical website (results in your browser may vary).

Figure 8-4. Facebook Header.

Note the "s" in https and the lock.

Figure 8-5. Facebook Header Marked.

The "s" stands for "secure." Facebook has published a public key. Your browser will take the password you enter and use the public key to encrypt it. The encrypted key is sent to Facebook. The man with a laptop on the other side of the coffee shop can't figure out your password even if his computer is scanning everything sent via Wi-Fi. Facebook can recover your password using its private key. Likewise, your browser will create its own private and public keys and send the public key to Facebook. That way Facebook can send you encrypted versions of the status updates of your friends which no one else can see.

Cryptography If P = NP

What would happen to cryptography if we were in the "beautiful world" of chapter 2 where P = NP? We could easily compute whether 5,754,853,343 and 2,860,486,313 were prime numbers and that 5,754,853,343 × 2,860,486,313 = 16,461,679,220,973,794,359. We could even do these computations for numbers that were thousands or millions of digits long. Since we would be able to verify the solution to a factoring problem, the factoring problem would sit in NP. If P = NP, then factoring would now be computationally efficient, and we could find the prime factors of numbers that are millions of digits long. P = NP would break the RSA protocol. P = NP breaks all public-key protocols, for if P = NP, you can always recover the private key from the public key. If P = NP, it is impossible to send secret messages to someone you have never communicated with before.

Does that mean no cryptography in the beautiful world? There is one scheme, the one-time pad, which is provably secure with no assumption about the hardness of NP problems. Suppose Alice had a twelve-character password, FIDDLESTICKS. The pad is a series of random letters of the same length, JXORMQNAMRHC. Take the first letter of both password and pad, F and J, the sixth and tenth letters of the alphabet. Add together their order to get 16, and use the sixteenth letter, P, as the first letter of the encryption. Now take the second letters, I and X, the ninth and twenty-fourth letters, and add together their order to get 33. There is no thirty-third letter of the alphabet, so subtract 26, which yields 7, and use the seventh letter of the alphabet, G, as the second letter in the encryption. Continuing in this fashion produces the encrypted message "PGSVYVGUVUSV." Alice sends "PGSVYVGUVUSV" to Facebook. Facebook simply subtracts instead of adding to decrypt this message using the pad.

Since there are as many pads of length twelve as there are messages, all encryptions are equally likely, and it is mathematically impossible to learn anything about the message given the pad, no matter whether or not P = NP. So why don't we all use one-time pads instead of the more complicated and possibly breakable schemes based on factoring numbers?

You have to use one-time pads very carefully. True to their name, one-time pads can only be used once. Ever. Even two separate people sending secret messages to two other separate people had better not use the same pad, or both their messages might be compromised. Each one-time pad needs to be as long as the message that is being encrypted. There is no separate public and private key, just a shared private key (the pad JXORMQNAMRHC). Both Facebook and Alice need to know the pad, but if an eavesdropper learns even a small amount of the pad, he learns some information about the message. Facebook has to get Alice the pad (or vice versa) without anyone else seeing it. Since people can listen in on Internet communications, Facebook or a trusted intermediary would have to send Alice a one-time pad on a USB drive or other physical device. I can imagine in the beautiful world Alice would buy a sealed USB drive full of one-time pads at a local store. The drive would be created by some trusted organization (the U.S. government?) with some method of providing the same pad securely to Facebook and other companies as well.

P = NP makes most things much easier, but it makes cryptography a bit more painful.

Facebook could also use quantum mechanics to both create and transmit a pad to Alice. We'll talk more about quantum cryptography in the next chapter, but that approach will likely be too expensive to use on a large scale.

Zero-Knowledge Sudoku

Bob was spending his lunch hour struggling to solve a Sudoku puzzle from today's newspaper.

Bob exclaims in agony, "There must be a mistake in the paper, for this Sudoku puzzle cannot have a solution." Alice, Bob's co-worker, overhears and takes a look at the puzzle. She had solved the same puzzle on the morning train and so knew that Bob was wrong. Here is Alice's solution.

	9			8		4		
		2		4	1			5
3							6	
	1							
7	6			2			1	9
							8	
	2							8
5			2	9		3		
		4		5			2	

1	9	7	6	8	5	4	3	2
6	8	2	3	4	1	7	9	5
3	4	5	9	7	2	8	6	1
4	1	8	5	6	9	2	7	3
7	6	3	8	2	4	5	1	9
2	5	9	7	1	3	6	8	4
9	2	6	4	3	7	1	5	8
5	7	1	2	9	8	3	4	6
8	3	4	1	5	6	9	2	7

Figure 8-6. Zero-Knowledge Sudoku.   **Figure 8-7.** Zero-Knowledge Sodoku Solved.

Bob is about to give up on the puzzle, so Alice tells him that she was able to solve the puzzle, but Bob doesn't believe her. Alice knows Bob will be really upset when he reads the solution in tomorrow's paper, so she needs Bob to keep working on the problem. Alice could convince Bob by showing him her solution, but this would destroy the fun of the puzzle for Bob. Alice wants to convince Bob that the Sudoku puzzle has a solution without revealing any information about that solution.

Luckily, Alice was a computer science major in college and knows about zero-knowledge proofs. She comes up with the following scheme.

First, Alice goes back to her cubicle, where Bob can't see her. Alice chooses a random reordering of the digits.

Old	New
1	2
2	8
3	6
4	5
5	4
6	9
7	1
8	7
9	3

Figure 8-8. Digits.

She then takes her solution and renumbers it according to the table above (replacing 1 with 2, 9 with 3, etc.), writing this new grid on a large piece of paper.

2	3	1	9	7	4	5	6	8
9	7	8	6	5	2	1	3	4
6	5	4	3	1	8	7	9	2
5	2	7	4	9	3	8	1	6
1	9	6	7	8	5	4	2	3
8	4	3	1	2	6	9	7	5
3	8	9	5	6	1	2	4	7
4	1	2	8	3	7	6	5	9
7	6	5	2	4	9	3	8	1

Figure 8-9. Zero-Knowledge Sodoku Renumbered.

Alice then carefully cuts up the paper into eighty-one pieces, each with one digit in it. She carefully takes each of these pieces and puts them into eighty-one little bags and arranges the bags with their numbers in their proper place on a blank grid.

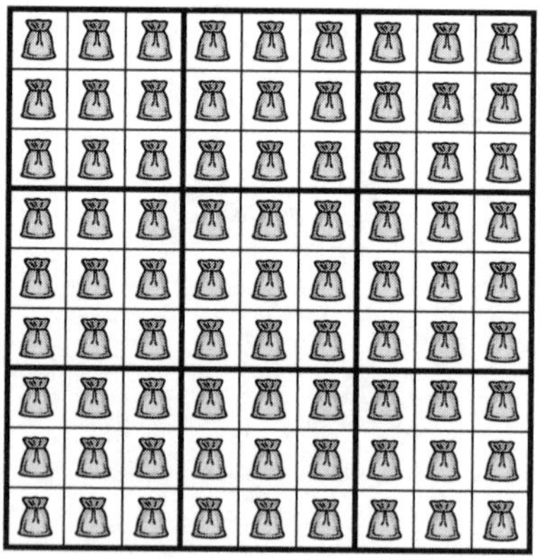

Figure 8-10. Zero-Knowledge Sodoku Covered.

The top left bag contains the number 2, the bag to the right of that contains the number 3, and so on.

Alice carefully takes the grid with the bags on them back to Bob. She explains what she has done (without revealing the new numbering scheme) and lets Bob choose one of twenty-eight options:

- Pick one of the nine rows, and open all the bags in that row.
- Pick one of the nine columns, and open all the bags in that column.
- Pick one of the nine 3 × 3 subsquares, and open all the bags in that subsquare.
- Choose the bags corresponding to the positions of the numbers of the original puzzle, and open those.

Suppose Bob chooses the third row and opens those bags. This is what he sees.

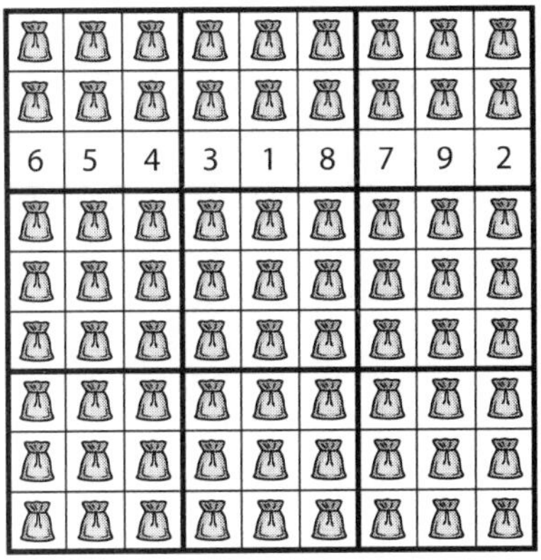

Figure 8-11. Zero-Knowledge Sodoku Row.

If Alice had a solution and did as she said, the numbers Bob sees should be all the digits from 1 to 9 in some random order. If Bob sees two of the same digit, he knows Alice has cheated, but here Alice passes the test.

Bob has similar tests if he opens a column or subsquare.

Suppose Bob chooses that last option, "Choose the bags corresponding to the positions of the numbers of the original puzzle, and open those." Here is what Bob sees after opening up the bags corresponding to the numbers in the original puzzle.

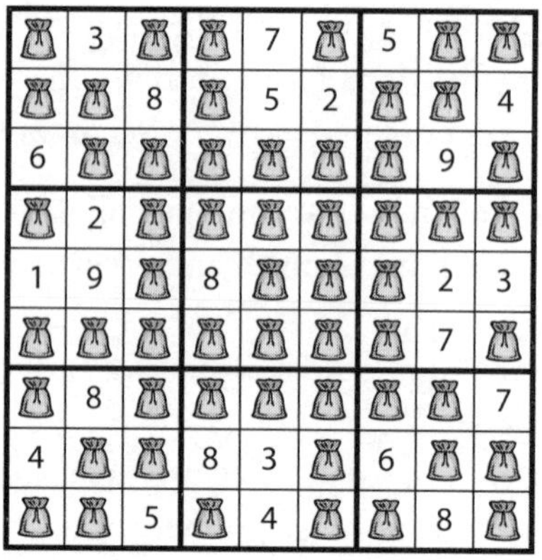

Figure 8-12. Zero-Knowledge Sudoku Corresponding.

In this case and only in this case, Alice also gives Bob the renumbering table.

Old	New
1	2
2	8
3	6
4	5
5	4
6	9
7	1
8	7
9	3

Figure 8-13. Digits.

Bob then checks that the numbers he has opened corresponds to the original puzzle according to the numbering scheme that Alice has provided. The number 9 in the original puzzle according to the scheme should be numbered 3 in the version with the bags, and indeed it is.

If Alice really does have a solution and followed all the rules as she describes them, then she will pass all the tests. What does Bob learn?

If Bob chooses a row, column, or subsquare, all Bob will see is the nine digits randomly placed in that row, column, or subsquare, giving him no useful information in solving the puzzle.

If Bob chooses the last option, he will see just a random permutation of the original puzzle, also giving him no useful information to solve the puzzle.

What if Alice doesn't really have a solution? No matter what Alice does, she will fail one of the tests that Bob could perform. If Bob chooses a test at random, Alice will be caught at least one out of twenty-eight times, or approximately 3.57 percent of the time. That's not a very high chance of being caught; however, if Alice and Bob repeat the whole procedure eighty-three times (choosing a new renumbering each time), Alice will fail a test one of these times with probability over 95 percent.

Bob becomes convinced that Alice really has a solution to the Sudoku puzzle, but Bob has received "zero knowledge" about that solution other than that a solution exists. Bob can now go back to work on the puzzle confident that a solution exists, but without any help in finding that solution.

In our example, Bob doesn't want to find out any details about the solution. But Bob can cheat by grabbing all the bags and opening them all up. Alice can prevent this by using locked boxes instead of bags and giving Bob only the relevant keys.

What if Bob and Alice are not in the same office but instead in different cities? Alice and Bob can still easily communicate by phone or email, but what can they do about the bags? They can use simple cryptography. Alice chooses a different random large number for each bag whose last digit matches the one in the bag, for example 3,682,502 for a bag that contains the number 2. Alice then encrypts each of these numbers using her public key and sends the encryptions to Bob. After Bob chooses his option as before, Alice gives the decrypted values only for those corresponding to the bags Bob is allowed to open. Bob can re-encrypt those messages using Alice's public key and check that they match Alice's original encryptions.

As we mentioned back in chapter 4, Sudoku is NP-complete, so every NP problem can be reduced to Sudoku. So we automatically get a zero-knowledge system for every NP problem. Alice can convince Bob that there is a large clique in Frenemy, a three-coloring of a map, or a short traveling salesman tour, all without revealing any information about the clique, coloring, or tour other than that they exist.

In a common attack in cryptography, a person can cheat by pretending to be someone he or she isn't. Zero-knowledge proofs give a method to guard against such identity spoofing. Alice encrypts some randomly chosen secret using her public key. Alice wants to convince Bob that she is indeed Alice. Alice could just give Bob the secret, but then Bob could pretend to be Alice. Instead, Alice gives a zero-knowledge proof convincing Bob that she knows the secret. So Bob is now convinced he's talking to Alice, but he doesn't learn anything about Alice's secret.

Playing Games

Bob and Alice are trying to decide where to eat dinner. Bob would like a steak place but Alice wants fish. They decide to flip a coin to choose the restaurant. So Bob flips the coin and covers it up on his arm. Alice calls heads. Bob reveals the coin to be tails and enjoys a fine steak that evening.

Sounds fair, but what if Bob and Alice are talking on the phone or over the Internet? Bob may lie about whether the coin is heads or tails. Maybe Bob doesn't even flip the coin at all. How can Alice be sure?

One method is to use public randomness. Bob and Alice could agree that if the last digit of the closing Dow Jones Industrial Average is odd, then Bob picks the restaurant, and if it is even then Alice gets to choose. But this doesn't work on a Saturday when the markets are closed.

Here is a cryptographic solution using a public-key scheme we saw earlier in this chapter. First Bob creates public and private keys. Then Bob picks a random number, say 69,441,251,920,931,124, and encrypts that number using the public key. Bob sends the public key and the encryption of 69,441,251,920,931,124 to Alice.

Alice then guesses whether Bob's number was even or odd and sends "even" or "odd" to Bob. Bob then sends the private key to Alice. Alice uses the private key to decrypt the encrypted number and determine that Bob's original number is 69,441,251,920,931,124. Had Alice said "even," then Alice would win the coin toss. If Alice had said "odd," then Bob would win.

Why does this scheme work? Bob chooses his number, 69,441,251,920,931,124, before Alice chooses even or odd. When Alice chooses even or odd she has no idea that Bob is using 69,441,251,920,931,124. All Alice sees is the encrypted value, and she can learn nothing about Bob's number unless she can break the code. So Alice can at best do a random guess. Bob can't change the number since he has already sent the encrypted version to Alice. So Bob reveals his original number by revealing the private key. If both players play randomly, each will win half the time. Neither person can cheat as long as public-key cryptography is secure.

A coin toss is pretty simple. How about more complicated games?

Can Bob and Alice play chess over the telephone? Quite easily: each one in turn just tells the other their next move using standard chess notation.

How about backgammon or Monopoly? How does Bob throw the dice and get Alice to believe the outcome? They use a dice-rolling protocol very similar to the coin-flipping protocol above.

How about poker or other card games? Now this gets much more complicated. Alice should get some randomly chosen cards that she can see but Bob can't see. Bob also should get some cards that Alice can't see. There might also be other cards that both players can see and cards that neither can see but that are revealed to one or both players later in the game.

There are many websites where you can play poker, but these sites act as a trusted party that deals the cards and reveals them to each player on the player's own browser.

Can Bob and Alice play poker without a trusted website? The coin-flipping scheme is not enough. In the 1970s and 1980s cryptographers came up with a variety of schemes for playing card games over the telephone or Internet that involve two or more players having private and public keys and doing encryptions of encryptions.

In the 1980s and 1990s cryptographers developed very general techniques whereby any game played with a trusted adversary can be played without a trusted adversary over the Internet. These methods use both encryptions and zero-knowledge proofs. These protocols are quite complicated and rarely used in practice. In the real world, people rely either on trusted websites or on protocols tailored for a specific purpose.

Secretly Computing in the Cloud

Suppose Alice needs computation done on sensitive data and Bob runs a cloud computing service. Alice can send her data to Bob using Bob's public key. Bob decrypts the data, does the computation, and sends the results back to Alice using Alice's public key. Assuming that Alice and Bob use a secure public-key protocol, no eavesdropper will see Alice's data. That's fine as long as Alice trusts Bob. What if Alice wants to keep her data secret even from Bob?

To solve this problem, we need something called fully homomorphic cryptography. In RSA, if you take the encrypted versions of two numbers, say 28 and 45, and multiply them together, you get the encryption of the product, 1,260. You don't need to know what the original numbers were but you can still compute the encryption of the product from the encryptions of the original numbers. On the other hand you can't do the same for sums; there is no known way to go from the RSA encryptions of 28 and 45 to the encryption of 73.

Most computations can be expressed by a combination of sums and products. Sums are similar to ORs and products are like ANDs, so you can build circuits with sums and products. A fully homomorphic encryption scheme lets you compute both the encrypted sums and products of encrypted numbers. Using such a scheme you can compute the encrypted output of a computation from the encrypted inputs without any additional communication.

Alice could encrypt her data using a fully homomorphic scheme and upload the encrypted information to Bob's computers. Bob could perform the calculations needed by Alice while remaining in the dark about Alice's data. At the end, Bob would have encrypted versions of the results of the computation. Bob himself could not decrypt those results but Alice could once she downloaded them from Bob's computer.

For many years, fully homomorphic encryption remained beyond the reach of cryptographers, and many thought it might be impossible to achieve. In 2009 Craig Gentry, a Stanford graduate student, developed a method for a fully homomorphic encryption scheme. His scheme is not yet efficient enough for practical use, but it certainly opens the door to much better cryptographic protocols in the very near future.

Creating Randomness

Rock-paper-scissors is a popular game in which two players choose a rock, paper, or scissors using hand signals.

Since rock can crush scissors, the player choosing rock beats the player choosing scissors. Scissors cuts paper, and paper covers rock. If both players play the same item, they tie.

What's the best way to play rock-paper-scissors? If you can predict your opponent well, then play accordingly. But let's consider the opposite situation, where your opponent knows your thought processes. Will you always lose?

Not if you choose at random. If you choose rock, paper, or scissors with equal probability, then no matter what your opponent does you will win a third of the time, lose a third of the time, and tie a third of the time. Humans are particularly bad at making perfectly random choices, which is why there is actual skill involved in rock-paper-scissors tournaments.

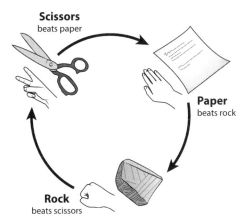

Figure 8-14. Rock-Paper-Scissors.

Similarly, cryptography also requires randomness. Every crypto-graphic protocol we have talked about in this chapter needed random choices to keep information secret from eavesdroppers and adversaries. If our numbers aren't completely random, the adversaries can gain an advantage, even for one-time pads.

How do we generate random numbers? Computers don't flip coins, and even if they did, coins still follow physical laws and do not generate true randomness. One might be able to extract randomness from quantum effects, but that's not likely to be practical for everyday computing.

To generate randomness, we humans flip coins, roll dice, shuffle cards, or spin a roulette wheel. All these operations follow direct physical laws, yet the casinos are in no risk of losing money. The complex interaction of a roulette ball with the wheel makes it computationally impossible to predict the outcome on any one spin, and each result is indistinguishable from random.

Computers take a similar tack. Computers don't have access to truly random bits; rather, they generate "pseudorandom" numbers by performing operations whose outcomes are hard to predict. There is a direct connection between cryptography and random number generation. The encryption of a secret should look random to an adversary trying to find the secret without the key. Many encryption techniques can be turned into good random number generators.

Your computer probably does not use these systems to create pseudorandom numbers, but uses more efficient systems that usually work well but don't have the theoretical guarantees. How we trade off computation time for better pseudorandom generators is a tricky choice to make.

Pseudorandom number generators work well if we have suitably complex problems. If P = NP, every efficiently computable process can be inverted in some sense. It would be hard, if not impossible, to create computationally random coin tosses. Rock-paper-scissors would not be a game of chance in the beautiful world.

The Continuing Challenges

We use public-key cryptography based on NP problems like factoring. Alice can easily generate hard-to-factor numbers where she knows the factors just by picking two large, randomly chosen prime numbers and multiplying them together.

Factoring is not known or even believed to be NP-complete. Even if factoring were NP-complete, P ≠ NP just means there would be some numbers that are hard to factor. It could be the case that randomly chosen numbers would still be easy to factor.

Basing cryptography on NP-complete problems and on a P ≠ NP assumption remains a difficult challenge in cryptography.

The change in cryptography started by Diffie and Hellman in the 1970s has allowed us to build cryptographic protocols based directly on the hardness of solving specific problems. No longer will crossword puzzle solvers, chess players, and smart mathematicians find ways to directly break these codes.

The cat-and-mouse game still goes on. Instead of breaking the codes directly, hackers work on other vulnerabilities. Alice and Bob might fail to use good random numbers in their protocols. A hacker could break into Alice's computer because of a design flaw in her browser or operating system. Alice might be tricked into revealing her private keys. Alice might make herself vulnerable because she didn't use a good password.

A hacker could use information beyond just the communication that he sees. Perhaps the amount of time a computation takes might vary depending on the message encoded. A hacker might damage part of the system, such as by sticking a smart card in a microwave, and hope the resulting damaged computation may no longer keep secrets.

We may have unbreakable codes, but building systems that will always keep secrets a secret may never come to be.

Chapter 9

— — — — — — — — —

QUANTUM

IN 1982, THE NOBEL PRIZE–WINNING PHYSICIST Richard Feynman noticed there was no simple way of simulating quantum physical systems using digital computers. He turned this problem into an opportunity—perhaps a computational device based on quantum mechanics could solve problems more efficiently than more traditional computers. In the decades that followed, computer scientists and physicists, often working together, showed in theory that quantum computers can solve certain problems, such as factoring numbers, much faster. Whether we can actually build large or even medium-scale working quantum computers and determine exactly what these computers can or cannot do still remain significant challenges. In this chapter we explore the power of quantum computing, as well as the related concepts of quantum cryptography and teleportation.

The Quantum DVR

Tom is a Bostonian and, of course, a huge Red Sox fan. The New York Yankees played in Boston earlier in the day but Tom had to work, and he purposely avoided reading anything about the game. When he got home he ordered in some pizza, fired up the DVR, and started watching the game hours after it had ended. In the bottom of the ninth inning the Red Sox had men on second and third with two out and one run down. The Boston slugger Brian Hammer walked up to the plate, and Tom hoped that Hammer would get a hit. Tom stopped himself, the game was long over. Hammer had either got a hit or he hadn't. But Tom didn't know which one. In Tom's view the outcome of the game hadn't been determined yet; the truth lay somewhere between win and loss, and soon would be revealed to him.

The truth for Tom lies in his observations. From Tom's point of view, until he observes the final play the game hasn't ended yet, the winner is yet to be determined. Before that final play, the game exists in some strange state between the Red Sox winning or losing.

Susan, a diehard Yankees fan, also recorded the game on a DVR and watched it play out at about the same time as Tom. Like Tom, Susan had no idea whether or not Hammer would get the hit causing the Yankees to lose. In Susan's mind the game was also not decided. It was still a random event until she saw the end.

Susan and Tom are observing two random events at the same time 200 miles apart from each other. However, they will see the same outcome: whether Hammer gets his hit or not will happen for both Tom and Susan or for neither. It is impossible for Tom to see Hammer get the hit and Susan to see him strike out. Neither knows the outcome, but both know they will see the same outcome when the game plays out. The results as they play out on Susan's and Tom's DVRs are somehow entangled with each other.

What does this have to do with quantum computing? Traditional digital computers have as their most basic element the bit (short for **b**inary dig**it**), something that can take on one of two values such as win or lose or true or false. The basic element of a quantum computer is called a qubit, for **qu**antum **bit**. While a bit can take on one of two values, a quantum bit can take on something in between.

Baseball games recorded to DVRs aren't really qubits, but they share some of the same properties. When Tom watches the ball game the outcome is not determined until the game ends. But when Tom observes the end of the game, the outcome is now determined: it is either win or lose. No longer is it in between. Likewise, when a qubit is observed, it also becomes a traditional bit, taking only one of two values, no longer something in between.

Quantum bits can be entangled, like the DVRs of Tom and Susan. We can have two quantum bits so that they always give the same answer when observed.

But that's where the analogy ends. Quantum bits can be entangled in far more complex ways, and these entanglements can be manipulated to create computation.

The outcome of the game, a simple probability, can take on values from a simple line.

Figure 9-1. Boston.

The star represents about a 30 percent probability of a Boston win. Once Tom watches the game, the star will move either all the way to the left or all the way to the right, depending on the outcome.

A qubit, on the other hand, takes on values in a circle.

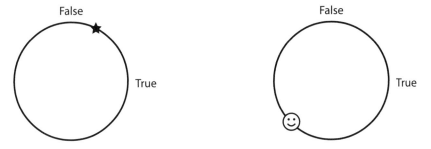

Figure 9-2. Qubits.

The star now has two dimensions sitting at (0.55 True, 0.84 False). A qubit can take on negative values as well. The Smiley face represents a qubit of values (–0.71 True, –0.71 False). A quantum computer can rotate and flip this circle around in a controlled way.

One quantum bit is represented by a circle, an object in two dimensions. Two quantum bits need a four-dimensional version of a circle to represent them, quite difficult to draw here or even visualize. Thirty quantum bits require over a trillion dimensions.

This suggests an approach to solving NP problems using a quantum computer. Think of trying to find a clique of size fifty among the 20,000 inhabitants of Frenemy. With about 500 qubits one can get a mixture of all groups of size fifty. We can process all these groups at the same time in parallel and mark the ones that form a clique by doing an appropriate set of rotating and flipping operations.

We now have a "quantum state" that is some combination of about 3×10^{150} (3 followed by 150 zeros) groups of Frenemy citizens, some of which are marked as cliques. If there were some quick way to pull out those cliques from the quantum state, we would have a quick quantum way to solve clique and every other NP problem. When we look at the quantum state (akin to watching the end of a game on a DVR) we only see one outcome, one of the groups of people in Frenemy, and most likely the group we see won't be a clique.

We need a way to make the groups that are a clique stand out and more likely to be seen when we view the state. We can do this with quantum manipulation. A naive approach would take as long as there are groups, roughly 3×10^{150} quantum manipulations, losing any advantage of using quantum computing. In 1996 Lov Grover, working at Bell Labs in New Jersey, developed a smarter quantum algorithm that would pull out the clique in "only" 2×10^{75} quantum operations. Even if we could do a trillion quantum operations a second this would still take about five times the current age of the universe.

There is some evidence that the Grover algorithm is the best one can do for NP-complete problems using a quantum computer, so it is unlikely that a quantum version of P = NP could be true. Even if physicists managed to build quantum computers, our hardest problems would remain beyond their reach.

That doesn't mean quantum computers would not be useful. They could do complex simulations of nano-scale physical systems efficiently that might help unlock some of the mysteries of the universe. Quantum computers could also solve some NP problems that we don't know how to solve efficiently on traditional machines.

In 1994 Peter Shor, also then at Bell Labs, realized that quantum computers could factor numbers, for example, taking the number 16,461,679,220,973,794,359 and determining the two numbers 5,754,853,343 and 2,860,486,313, where 5,754,853,343 × 2,860,486,313

= 16,461,679,220,973,794,359. His algorithm would work on a fast quantum computer even for numbers that were hundreds or thousands of digits long. Searching for factors of a number has an algebraic structure that quantum computers could take advantage of. While we think factoring is very difficult for today's computers, a quantum computer could overcome these difficulties to factor large numbers. NP-complete problems lack a nice algebraic structure, which is why Shor's algorithm won't work to solve general NP problems.

Of course, before we can use Grover's or Shor's algorithms, we need a working quantum computer. To solve reasonably sized problems that today's machines cannot solve we'll need at least tens of thousands of quantum bits entangled and manipulated for several seconds. Unfortunately, entanglement is quite fragile. Any interaction with a quantum system and the environment can cause an "observation" that can lead to some loss of entanglement, which can wreak havoc with a delicate quantum computation.

Physicists have yet to achieve perfect or even near-perfect entanglement on even two quantum bits. Computer scientists have developed methods known as quantum error-correction to develop algorithms that can handle a moderate amount of entanglement. Even so, we do not know how to create a significant amount of entanglement in more than a handful of quantum bits. It might be some fundamental rule of nature that prevents significant entanglement for any reasonable length of time. Or it could just be a tricky engineering problem. We'll have to let the physicists sort that out.

There are other ways to create computing via quantum effects through techniques called quantum adiabatic systems or quantum annealing, though they have their own technical and computational limitations. A company known as D-Wave claims to have developed machines based on these techniques, but the jury is still out on whether they can compute beyond our desktop machines.

Even if we discover how to build full-scale quantum computers they will remain special-purpose machines, useful for factoring numbers or simulating quantum systems. They may help us break codes and help us better understand the fundamental nature of the universe, but they likely won't solve NP-complete problems or make spreadsheets run faster.

Quantum Cryptography

Most of the cryptographic tools we discussed in chapter 8 are secure under the assumption that factoring is a difficult computational problem. Anyone with a quantum computer in her pocket could break these codes by using Shor's algorithm to factor the keys. We don't yet have quantum computers in our pockets, but that could happen someday. Cryptographers could try to get around quantum attacks by developing protocols based on other hard problems that lack the algebraic structure that quantum computers could exploit. The quantum community offers us another solution, cryptography based on quantum mechanics.

We take it for granted that we can make copies of computer data. Copy and Paste functions exist on nearly every computing platform. We can save our files in multiple directories or on different machines. We run backups of our data onto hard drives or in the "cloud." If anything, we have a problem with too many copies of our information. It's hard to delete our personal files or email and make sure no copies remain somewhere.

Quantum bits, on the other hand, cannot be copied. To copy a quantum bit you need to look at it, at least in a limited way, and viewing a quantum bit turns that bit into a traditional true-false bit. George can send Harry a quantum bit. If Eric is on the line and tries to copy or read that bit he will change the bit to a traditional bit. This alone does not give a way to send secret messages; after all, Harry will also have to read that bit which also changes it.

Gilles Brassard from the University of Montreal attended one of the major meetings in theoretical computer science, the 1979 IEEE Foundations of Computer Science conference held that year in Puerto Rico. Gilles was swimming at the beach when Charlie Bennett from IBM tracked him down. That meeting in the ocean led to an amazing collaboration in the course of which they discovered a method to use quantum bits to create a provably unbreakable cryptographic system. George would send Harry a large number of quantum bits encoded in various ways. If an enemy, say, Esther, tried to intercept these quantum bits, either to view them or attempt to copy them, she would alter those bits. Bennett and Brassard developed a method that would allow George and Harry, with some additional communication, both classical and quantum, either to successfully transfer a secret key they could use for future encoding or to detect that the key was compromised, in which case they could try again.

The protocol is not without limitations. Small errors both increase the number of quantum bits needed as well as open up the possibility that Esther could do some cheating and remain undetected. A series of more sophisticated protocols have been developed to handle those challenges.

Unlike quantum computing, the basic Bennett-Brassard protocol does not require quantum entanglement, and also unlike quantum computing, there are working implementations of medium-scale quantum cryptography. Researchers at Los Alamos have successfully sent messages over ninety-two miles through fiber optic cables. Others have sent messages between two of the Canary Islands through the air, a distance of nearly ninety miles. One day in the near future we may be able to use quantum techniques to send unbreakable codes via satellite.

So why don't we use quantum cryptography for all secrecy needs? Quantum cryptography is still in the experimental stage, quite expensive, not much bandwidth, and prone to error. Vulnerabilities in cryptographic systems usually don't rely on breaking the underlying codes but weaknesses in the implementation. It's just as easy, probably much easier, to have as bad an implementation of a quantum protocol as of a classical one. There is no clear way to do quantum cryptography through the Internet, where information is routed through many stations from source to destination. And most important, we won't likely see any computer, quantum or otherwise, that can factor numbers in the near future. The current codes will remain secure for quite a while.

Quantum Teleportation

On the inside front cover of the February 1996 *Scientific American*, IBM ran a two-page ad trumpeting new research:

> For years she shared recipes with her friend in Osaka. She showed him hundreds of ways to use paprika. He shared his secret recipe for sukiyaki. One day Margit e-mailed Seiji, "Stand by. I'll teleport you some goulash." Margit is a little premature, but we are working on it. An IBM scientist and his colleagues have discovered a way to make an object disintegrate in one place

Figure 9-3. IBM Ad. Courtesy of IBM.

and reappear intact in another. It sounds like magic. But their breakthrough could affect everything from the future of computers to our knowledge of the cosmos. Smart guys. But none of them can stuff a cabbage. Yet.

What was this amazing process trumpeted by the ad? It had little to do with goulash than with quantum bits. But a quantum bit can really disintegrate and reappear somewhere else. And it does seem like magic. The ad refers to 1993 research from IBM scientist Charlie Bennett and his colleagues, including Gilles Brassard. The people who gave us quantum cryptography also gave us quantum teleportation.

Arthur has a quantum bit and wants to give it to Harriet. How does he do that?

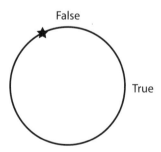

Figure 9-4. Qubit.

Arthur could just send the qubit via Federal Express, but if a customs official decides to inspect the contents and observes the qubit, the quantum bit is destroyed, replaced by only true or false. Even if Arthur uses great care to transport the bit to Harriet, it is nearly impossible for him to avoid interaction with the outside environment that could damage the qubit.

Arthur could send the description of the qubit (−0.55 True, 0.84 False) and Harriet could reconstruct an identical qubit. This only works if Arthur already knows the details of the qubit. If Arthur has just the qubit, he cannot determine its exact description, for any attempt to measure the quantum bit collapses it to just true or false.

Bennett and his colleagues developed a method for Arthur to transmit his qubit to Harriet, but with a caveat. Arthur and Harriet need previously entangled quantum bits.

The qubits with the heart that Arthur and Harriet have are fully entangled. If Arthur were to look at his qubit, he would see "True" or

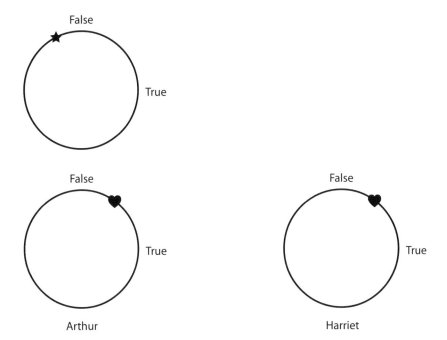

Figure 9-5. Entangled.

"False" with equal probability. But if Arthur sees "True," then Harriet sees "True," and if Arthur sees "False," then Harriet also sees "False," similar to the baseball game from earlier in the chapter.

Arthur now has two qubits, the one he wants to send (the star) and the one entangled with Harriet (the heart). These two qubits combined sit on a four-dimensional sphere, as we mentioned earlier. Without looking at the qubits Arthur performs a careful rotation of this sphere in four dimensions. Arthur now measures both qubits to get two traditional classical bits (zeros and ones).

Arthur sends these two classical bits to Harriet. One of these bits tells Harriet whether to rotate her qubit, the other whether to flip it. Harriet performs these operations as prescribed. Presto! Mumbo! Nothing up her sleeve! Harriet now has in her possession Arthur's original qubit. Shockingly, this all works.

What happened to Arthur's qubits? They were destroyed when they were measured. Otherwise Arthur would have successfully created a copy of his qubit, something ruled out by the laws of quantum mechanics.

While Arthur is manipulating and measuring his qubits, Harriet's qubit remains the same. Arthur cannot do anything to his qubits that affects Harriet's qubit, even though Harriet's qubit is entangled with one of Arthur's. Otherwise Arthur would have a way of communicating with Harriet, providing instantaneously information that can travel faster than the speed of light, a big physics no-no.

When Harriet receives those two bits from Arthur, this information allows Harriet to view her qubit now as a twisted version of Arthur's original qubit. Those bits tell Harriet exactly how to untwist and recover

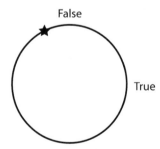

Figure 9-6. Qubit.

Arthur's qubit. Harriet cannot detect Arthur's manipulations, but when Arthur transmits his two measly bits, a whole quantum bit gets magically transmitted.

What does this have to do with goulash or those wonderful transporters from *Star Trek*? How would a transporter based on quantum teleportation work? Suppose I wanted to teleport from Chicago to Tokyo. Some company would create a large number of entangled qubits and carefully ship one side of the entangled qubits to Chicago and the others to Tokyo. I would then be placed in a chamber in Chicago with enough qubits to describe me, and together we would get rotated through some high dimensional space and get measured into classical bits. Those bits would be transmitted to the company's Tokyo division, which would do the appropriate rotations to the qubits in its chamber, open it up, and there I'd be. Cool.

Those entangled qubits are now spent. They would need new entangled bits to transmit someone else over or if I had to go back. On the other hand, we can say the same thing for jet fuel. I can't reuse the fuel needed to fly to Tokyo to get home.

More worrisome is maintaining full entanglement of the billion billion billion quantum bits needed to describe me (roughly the number of atoms in a human body), and that's probably a huge underestimate of how many qubits would be needed. Lose even a small amount of entanglement that comes from any slight interaction of the quantum bits with the outside environment and I would come out of the process seriously deformed and most likely a dead blob. So, you go first. I'll waste some jet fuel and meet you there.

Future of Quantum

Some researchers feel that we should rethink all of our computing in terms of quantum. Even if they don't solve NP-complete problems, their ability to simulate the quantum world will lead to great new advances in our understanding of matter, the cosmos, and even the brain, leading to scientific advances we cannot even imagine today.

Others see the very slow progress in quantum in both algorithms and hardware since the mid-1990s and wonder when, if at all, we will see quantum play a major role in computing devices. Short of a revolution, quantum computing will remain science fiction for some time to come.

If quantum is not the next wave of computation, what is? What great computational challenges await us? Read on.

Chapter 10

— — — — — — — — —

THE FUTURE

I've resigned myself to a bleak outlook for the P versus NP problem: I expect that P ≠ NP and that I will not see a proof in my lifetime. We won't see the beautiful world of chapter 2, but neither can we rule it out. The P versus NP problem will remain a mystery for decades and possibly centuries to come.

The P versus NP problem is more than a mathematical oddity. Even if we can't directly solve it, the P versus NP question gives us a common framework to think about and tackle the great computational tasks we need to solve. What are some of today's great challenges of computing?

- Parallel computation: Computers used to double in speed every eighteen to twenty-four months, but now we are hitting physical limits that prevent much faster processors in the future. Instead, computers are multiplying, and we can have several processors working together on a chip or in a cloud. How can we adjust our algorithms to work in this growing parallel world?
- Big data: From the Internet to scientific experiments and simulations, we create huge amounts of data every day. How do we reason about, make sense of, learn from, and develop predictions from this massive amount of information?
- Networking of everything: Much of the world lives on some computer network, whether they use some social network like

Facebook or just communicate via email. Soon nearly everything man-made will also be part of this network, from the
clothes we wear to the light bulbs we read by. How do we make
the best use of this ultraconnected world?

Whether or not it turns out that P = NP or P ≠ NP, thinking about P
and NP shapes how we address these challenges.

Parallel Computing

In 1965, Gordon Moore noticed that the number of basic components,
called transistors, on a computer chip greatly increased each year. Moore
made a bold statement that the number of transistors on a single chip
would double roughly every two years, and that that trend should hold
for another ten years. This principle, now dubbed Moore's law, continued well past 1975 through the present day and will continue for years
to come.

For a long while, Moore's law meant faster computing speeds. Around
2005, give or take a few years, computers started to hit physical limits
where the amount of energy needed to make processors go faster ex-
ceeded the advantages of faster speeds. If anything, processor speeds
have slowed down slightly to improve energy consumption.

Yet we still put more transistors on a chip. What do these transistors
do? Now these chips do more than one computation at a time, working
together in parallel, to solve problems faster by doing more than one
thing at a time.

Let's look at one machine in particular, IBM's Watson, which played
and won on the game show *Jeopardy* on episodes first broadcast in February 2011. Watson consisted of ninety IBM POWER 750 servers, each
with four POWER7 processors. A POWER7 processor actually has
eight processors (called cores) so it can run eight simultaneous computations at once. That's thirty-two cores per server for a total of 2,880
cores in the entire Watson system. Watson does 2,880 simultaneous
computations so it can interpret the *Jeopardy* "answer" and determine
whether to buzz in that fraction of a second before its opponents. The

2,880 simultaneous cores will seem a tiny number in the near future as Moore's law continues to put more on each machine. Over the next few decades we may have computers with millions or billions of cores.

The growth in parallelism will require techniques from across computer science. How can a computer decide to spread computation across multiple cores or computers and get the best performance possible? Do we need to rewrite our common programming languages to take into account multicore computers, and if so, what is the best method to do so?

The P versus NP problem gets impacted by parallelism. Remember Veruca Salt from *Charlie and the Chocolate Factory* back at the beginning of the book? She wanted a golden ticket that was hidden in some chocolate bar. Her father employed parallelism by dividing up the chocolate bars over the hundred workers he had at his peanut-shelling plant. One can do the same for NP search problems, dividing up the possible cliques among various computers and cores within the computers. This might turn a search that would have taken weeks into one that could take hours, but for some NP problems it may not help much. If we used a million computers, each with a billion cores, where each core can compute a quintillion operations a second, it would still take nearly a googol ages of the universe to look through all the possible cliques of fifty people among the 20,000 residents of Frenemy (a googol is 1 followed by 100 zeros). The P versus NP problem still remains relevant in a parallel world.

How about those problems in P, those we can solve efficiently? Can we take full advantage of multiple computers and cores? In most cases we can modify the algorithms to do exactly that. From arithmetic to finding the shortest path to matchmaking, all these problems have algorithms that can be spread over many cores.

Like P and NP, we have a name for the problems that computers can solve quickly in parallel, NC. NC stands for Nick's Class, after Nicholas Pippenger, one of the pioneers of parallel algorithms. If P = NC, then every problem we can efficiently solve, we can solve much faster using parallel machines. We don't know if P = NC, we don't even know if NP = NC. NP = NC means every NP search problem can be solved extremely quickly on systems with many computers and/or cores, an even more beautiful world than that of P = NP. It is highly unlikely that NP = NC, but it remains as big a mystery as P = NP.

Dealing with Big Data

In one second we create thirty-five minutes of YouTube videos, 1,600 Tweets, 11,000 Facebook posts, 50,000 Google searches, and three million emails (though 90 percent of them are spam).

The Hubble telescope scans the cosmos from its orbit, sending down 200,000 bytes of data every second. A byte of data is about one alphabet symbol. The Hubble's planned successor, the James Webb space telescope, basically a large parabolic mirror, will transmit to Earth up to 3.5 million bytes per second.

The Large Hadron Collider lies along the French-Swiss border and is the world's largest particle accelerator. On average it generates half a billion bytes of data every second. That's every second of every year, and there are 31 million seconds every year.

CERN, the European Organization for Nuclear Research that built and maintains the Large Hadron Collider, created a LHC Computing Grid, distributing its massive data to servers in thirty-four countries and allowing scientists worldwide to access and analyze that data.

A human DNA sequence takes about 55 million bytes to store. Sequences of all the seven billion people in the world would take nearly 400 quadrillion bytes. That's just the human beings.

We can easily and cheaply build sensors that measure everything from temperature to motion to sound to radiation. These sensors each constantly generate information and often one can employ thousands of sensors in the field. The U.S. Army is "swimming in sensors and drowning in data."

Not all data come from the outside world. Many scientific experiments are simply too difficult or too expensive to perform, so instead we rely on computer simulations to understand interactions in biological, physical, and chemical systems. These simulations again produce huge amounts of data waiting to be analyzed.

The data that come out of the above contain mostly useless information, either random noise or redundant data. Finding the crucial useful part of a data collection is difficult, and then we need to interpret that information. If P = NP, we would have algorithms that could work their way through the data, working through the important parts and using tools based on the principle of Occam's razor to help us understand and predict future events.

Since we likely don't live in that beautiful world, we have to find and tailor algorithms to serve the purposes we need. Making sense of large amounts of data is a difficult but very important computational task.

Big data can often be a boon, particularly in the area of machine learning. Machine learning works by training algorithms on data sets. The more data you have, the better your algorithm. Usually having more data trumps finding a better algorithm. Google does a reasonable job with spam detection, voice recognition, and language translation because it has large numbers of examples to work with.

In the near future, we will have data that should help us better analyze individual health, create smart grids that use electricity more efficiently, drive cars autonomously, and lead to new understandings of the basic nature of our universe. How we understand data to enrich our lives is a great challenge for computer scientists.

The Networking of Everything

Roughly two billion people are connected through the Internet in some way, via email or social networks. The Internet has allowed us to communicate, collaborate, learn, and play in ways unimaginable in the twentieth century.

What happens when we start putting things on the Internet? In the near future we will have small low-cost chips that can access the Internet via Wi-Fi, the cellular network, or one of many other wireless systems in development. We can put these chips on just about anything, from the clothes we wear to individual components in our cars to every package of food we buy. You can track everything, from whether your kids wear their seatbelts to your daily caloric intake, without having to write anything down. Running out of milk or shampoo? No worries: new supplies are automatically being provided and are on the way to your house. You won't take the wrong medications by mistake. The clothes you put on will tell you whether they are appropriate for the events and weather for the day or say, "Do you really want to wear this shirt with those pants?" This will be a great help to those who are colorblind or just have bad fashion sense.

You'll never lose your keys, wallet, tickets, or anything else ever again. You won't even have keys, a wallet, or tickets. Locks will open as you approach them through a signal from your phone. The ATM will just hand you money. You can just grab something at a store and walk out with it. Payments and deductions will happen automatically.

All these devices need to play nicely with each other. We need to develop the right methods that allow these devices to talk to each other but still maintain everybody's privacy. We could have coordination at an unprecedented level. Imagine how well traffic could move if all the cars worked together. These advances require solving large-scale problems quickly, reliably, and constantly. A small traffic accident can turn into large gridlock unless the system reacts rapidly. Some of these problems hit on the P versus NP problem, and we just have to deal with them as best we can.

Sun Microsystems (now part of Oracle) had a tagline in the 1990s: "The Network is the Computer." A network of machines, each acting individually, also works together as a single computational device. When everything is networked, we have a very large computer. Taming that machine won't be easy but will open the doors to great new possibilities.

Dealing with Technological Change

Parallel computation, big data, and networking everything aren't the stuff of science fiction. Rather, they are the manifestations of a sweeping change that has already started and will become pervasive within the next decade or two. How will society change?

We can't predict how technological innovations will change society. In the 1950s trucking company owner Malcolm McLean wanted to create a shipping container that could be moved directly from a ship to a truck without having to be unloaded. His invention led to a revolution in transportation, from huge ships designed to hold thousands of containers to automated ports that mostly eliminated the dock workers made famous in movies and TV shows like *On the Waterfront* and *All in the Family*. It also led to the rise of China, as the country could now ship its goods cheaply around the world—all from a steel box.

The automobile led to the rise of suburbia. Cell phones meant we no longer had to plan in advance. Twitter helped topple governments. Technologies change us in unpredicted ways. Get ready for it to happen again.

Beyond the computation issues, we need to incorporate human factors. People will do bad things with this new technology, both unintentionally and intentionally. We need good intuitive design to ensure that devices follow the wishes of users without much of a learning curve.

Some people may try to control technology, for personal gain or to learn information about other people. In the worst case people will use technology to cause havoc. Such havoc could be simply annoying, or it could be very costly and, in the worst case, deadly. Cryptography, security, and keeping a watchful eye will be needed to prevent others from causing major social distress.

There is always a point when technology is at its most dangerous: when it works well most of the time, so that people think it works well all of the time. We put much trust in a technology, and its failures catch us off-guard. Rare unexpected events can exacerbate this problem. Some examples are the failed levees in New Orleans during Hurricane Katrina, the massive oil spill that followed the explosion of the Deepwater Horizon drilling platform in 2010, and the disaster at the Fukushima nuclear power station after the 2011 Japanese earthquake and tsunami. We need to treat technology like the untamed beast it is. An unlikely but possible massive failure shouldn't lead to a colossal disaster.

Last Words on P and NP

Proving P ≠ NP will not be easy. You need to show no efficient algorithm, written now or in the future, can solve clique or any other NP-complete problem. How do you show every potential algorithm must fail?

It will happen, but it might take twenty or two hundred or two thousand years. Eventually we will develop new techniques that will finally prove P ≠ NP. When it happens, mathematicians will rejoice and hail it as a great solution to a great problem. The techniques developed to resolve the P versus NP question will give us great insights into the power of efficient computation, a concept that will permeate every aspect of our society.

P versus NP goes well beyond a simple mathematical puzzle. The P versus NP problem is a way of thinking, a way to classify computational problems by their inherent difficulty. Even without a proof that P ≠ NP, when we find ourselves needing to solve an NP-complete problem, we know we won't succeed in finding an algorithm that quickly solves the problem all the time. We need to rely on other tools, a combination of approximation, heuristics, and computational firepower, simply to do the best we can. NP-completeness gives us a common framework and allows us to create a toolbox of techniques that we can throw at these difficult-to-compute problems.

P versus NP brings communities together. We have NP-complete problems in physics, biology, economics, and many other fields. Physicists and economists work on very different problems, but they share a commonality that can give great benefits from sharing tools and techniques. Tools developed to find the ground state of a physical system can help find equilibrium behavior in a complex economic environment.

The inherent difficulty of NP problems leads to new technologies. Modern cryptographers have used P versus NP as an inspiration, taking cryptographic protocols from an art to a science. The need to solve NP problems has pushed us to create faster and more powerful computing systems, and helps push embryonic technologies like quantum computing.

Computation is about process, and not just on computers. The P versus NP problem has to do with the limits on nature itself, on how biological and physical systems evolve, and even on the power of our own minds. As long as P versus NP remains a mystery we do not know what we cannot do, and that's liberating.

- - - - *Acknowledgments* - - - -

FIRST I WOULD LIKE TO THANK Moshe Vardi, who encouraged me to write and served as editor for my *Communications of the ACM* review article, "The Status of the P versus NP Problem." The popularity of that article inspired me to expand it into a book for a broader audience.

My co-blogger Bill Gasarch encouraged me to keep writing and carefully reviewed the early drafts of all the chapters in manuscript form. Alana Lidawer and John, Jim, and Chris Purtilo similarly read early drafts of the entire book and offered many valuable comments. Kuan-Ling Chen, Josh Grochow, Ralph Hansen, Adam Kalinich, David Pennock, and Rahul Santhanam provided useful comments on early drafts of discrete chapters in the book.

Manuel Blum, Steve Cook, David Johnson, Leonid Levin, and Albert Meyer gave me many personal insights into the early days of P and NP. Alexander Razborov was very helpful with the Russian history.

This book draws on my life as a computer scientist, a life shaped by interactions with fellow researchers, students, and others too many to mention. Thanks are owed especially to my colleagues at the University of California at Berkeley, the Massachusetts Institute of Technology, the University of Chicago, the Centrum voor Wiskunde en Informatica in Amsterdam, the NEC Research Institute, the Toyota Technological Institute at Chicago, and Northwestern University for many valuable friendships and discussions.

I'd like to point out the two people who had the greatest early influence on my view of P versus NP: Juris Hartmanis, who first taught me about P and NP during my undergraduate years at Cornell, and Michael Sipser, my PhD adviser and mentor, at UC Berkeley and MIT.

For the map coloring examples in chapter 6, I turned to the web for help, and I thank those who answered my call: Chris Bogart, Hsien-Chih Chang, Dömötör Pálvölgyi, David Eppstein, Lukasz Grabowski, Gil Kalai, Charles Martel, and Derrick Stolee.

I wrote this book while a professor of electrical engineering and computer science at the Robert R. McCormick School of Engineering and Applied Science at Northwestern University. Northwestern truly encourages book projects that spread knowledge to the public. I made considerable use of Northwestern's facilities, particularly its extensive library collection, both physical and digital. The staff at Northwestern is top notch, and my administrative assistant, Marjorie Reyes, was especially helpful.

My editor at Princeton, Vickie Kearn, provided thoughtful guidance and a careful review of the manuscript at many stages, which has made for a much better book. I also would like to thank Vickie's assistant, Quinn Fusting, and the rest of the Princeton University Press staff.

The greatest thanks are owed to my family, my wife, Marcy, and my daughters, Annie and Molly, for their love and encouragement.

- - - - *Chapter Notes and Sources* - - - -

THE MATERIAL IN THIS BOOK was written based on my research experience in computational complexity and my interactions with thousands of other academic and industrial researchers who share an interest in the P versus NP problem. Some of this book was adapted from my blog, *Computational Complexity*.

In writing this book, I drew on a number of specific sources for stories, examples, and results. These sources are listed below.

Any updates to these sources or links, or significant errors discovered in the text, will be posted on the book's website http://press.princeton .edu/titles/9937.html. This website also has links to the cited books and talks, additional information, and further readings on the P versus NP question.

Preface

Lance Fortnow, "The Status of the P versus NP Problem," *Communications of the ACM* 52, no. 9 (September 2009): 78–86.
Stephen Hawking, *A Brief History of Time: From the Big Bang to Black Holes* (New York: Bantam Dell, 1988).

Chapter 1

The story of Veronica Salt is taken from Roald Dahl, *Charlie and the Chocolate Factory* (New York: Knopf, 1964).
The discussion of Yoku Matsuoka's research on an anatomically correct testbed hand incorporates information presented at a talk given at the 2010 CRA Snowbird Conference on July 18, 2010.
The traveling salesman problems were generated using software from Mark Daskin, http://sitemaker.umich.edu/msdaskin/software.

Chapter 2

Nearly everything in this chapter, except the section on Occam's razor, is a figment of my imagination meant to illustrate the unlikely world of P = NP.

Chapter 3

On Milgram's experiment, see Stanley Milgram, "The Small World Problem," *Psychology Today* 2, no. 1 (1967): 60–67.

The Bacon number calculation is from the Internet Movie Database.

For a readable story of the four-color problem, see Robin Wilson, *Four Colors Suffice: How the Map Problem Was Solved* (Princeton, NJ: Princeton University Press, 2004).

Chapter 4

The quotation from Cook is actually a paraphrase in modern terminology of the original quotation from his seminal paper. The original reads as follows:

> *The theorems suggest that {tautologies} is a good candidate for an interesting set not in L*, and I feel it is worth spending considerable effort trying to prove this conjecture. Such a proof would be a major breakthrough in complexity theory.*

Steve Cook, "The Complexity of Theorem-Proving Procedures," in *Proceedings of the Third Annual ACM Symposium on Theory of Computing* (New York: ACM), 151–58.

Karp's follow-up paper is Richard Karp, "Reducibility among Combinatorial Problems," *Complexity of Computer Computations* 40, no. 4 (1972): 85–103. According to Karp, "the reduction to CLIQUE is implicit in Cook's paper, and was also known to Ray Reiter."

Bob Sehlinger (author) and Len Testa (contributor), *The Unofficial Guide Walt Disney World 2010* (New York: Wiley, 2010).

"What's in a Name?" section is taken from Donald Knuth, "A Terminological Proposal," *ACM SIGACT News* 6, no. 1 (January 1974): 12–18.

Kevin Sack, "60 Lives, 30 Kidneys, All Linked," *New York Times*, February 19, 2012, A1.

Chapter 5

This chapter draws heavily from the following sources:

Lance Fortnow and Steve Homer, "A Short History of Computational Complexity," *Bulletin of the European Association for Theoretical Computer Science* 80 (June 2003), "Computational Complexity" column, and from personal discussions with a number of researchers, including Stephen Cook and Leonid Levin.

Dennis Shasha and Cathy Lazere, "A Good Solution Is Hard to Find," in *Out of Their Minds: The Lives and Discoveries of 15 Great Computer Scientists* (New York: Springer, 1995).

Juris Hartmanis, "Observations about the Development of Theoretical Computer Science," *Annals of the History of Computing* 3, no. 1 (January 1981): 42–51.

B. A. Trakhtenbrot, "A Survey of Russian Approaches to *Perebor* (Brute-Force Search) Algorithms," *Annals of the History of Computing* 6, no. 4 (October 1984): 384–400.

Michael Sipser, "The History and Status of the P versus NP Question," in *Proceedings of the 24th Annual ACM Symposium on Theory of Computing* (New York: ACM, 1992), 603–18. This article includes a copy of Gödel's letter to von Neumann and its English translation.

The story of Kolmogorov's attempt at history was confirmed by several Russian participants of the Centennial Seminar on Kolmogorov Complexity and Applications, held in Dagstuhl, Germany, April 27–May 2, 2003, and recorded in my blog, *Computational Complexity*, May 1, 2003.
Kolmogorov's story about saving probability theory in Russia comes from discussions with Alexander Razborov and a blog post (http://ansobol.livejournal.com/12551.html). As the text indicates, this story should perhaps be considered apocryphal.

Works Cited

Alan Cobham, "The Intrinsic Computational Difficulty of Functions," in *Proceedings of the 1964 International Congress for Logic, Methodology, and Philosophy of Science*, 24–30.

Stephen Cook, "The Complexity of Theorem-Proving Procedures," in *Proceedings of the Third Annual ACM Symposium on Theory of Computing* (New York: ACM, 1971), 151–58.

Jack Edmonds, "Paths, Trees and Flowers," *Canadian Journal of Mathematics* 17 (1965): 449–67.

Juris Hartmanis and Richard Stearns, "On the Computational Complexity of Algorithms," *Transactions of the American Mathematical Society* 117 (1965): 385–406.

Richard Karp, "Reducibility among Combinatorial Problems," *Complexity of Computer Computations* 40, no. 4 (1972): 85–103.

Leonid Levin, "Universal Sequential Search Problems" [in Russian], *Problemy Pred. Informatsii* 9, no. 3 (1971): 265–66. Translation in B. A. Trakhtenbrot, "A Survey of Russian Approaches to Prebor (Brute-Force Search) Algorithms," Annals of the History of Computing 6, no. 4 (October 1984): 384–400.

Warren McCulloch and Walter Pitts, "A Logical Calculus of the Ideas Immanent in Nervous Activity," *Bulletin of Mathematical Biology* 5, no. 4 (1943): 115–33.

Panel discussion, *Complexity of Computer Computations* 40, no. 4 (1972): 169–85.

Alan Turing, "On Computable Numbers, with an Application to the *Entscheidungsproblem*," *Proceedings of the London Mathematical Society* 42 (1936): 230–65.

S. Yablonsky, "On the Impossibility of Eliminating PEREBOR in Solving Some Problems of Circuit Theory," *Doklady Akademii Nauk SSSR* 124 (1959): 44–47.

Y. Zhuravlev, "On the Impossibility of Constructing Minimal Disjunctive Normal Forms for Boolean Functions by Algorithms of a Certain Class, *Doklady Akademii Nauk SSSR* 132 (1960): 504–6.

Chapter 6

The traveling salesman example is from "CRPC Researchers Solve Traveling Salesman Problem for Record-Breaking 13,509 Cities," a 2003 press release from the Center for Research on Parallel Computation at Rice University.

For help with the heuristic for map coloring and examples, I asked the community through a post at a Q&A site (http://cstheory.stackexchange.com/questions/4027/coloring-planar-graphs) and on my blog.

The province map of Frenemy is based on constructions in David P. Dailey, "Uniqueness of Colorability and Colorability of Planar 4-Regular Graphs Are NP-Complete," *Discrete Mathematics* 30 (1980): 289–93.

Chapter 7

The quotation from Juris Hartmanis in the first sentence is from a course he taught at Cornell University in spring 1985.

For the P versus NP policy, see the *Journal of the ACM* at http://jacm.acm.org/instructions/pnp.

Vinay Deolalikar's paper on P versus NP was attached to an email communication sent August 6, 2010, from Deolalikar to me and twenty-one other researchers.

The Cordano story is recounted in David Kahn, *The Codebreakers: The Story of Secret Writing* (New York: Macmillan, 1967).

Chapter 8

My account of the early history of cryptography draws heavily on David Kahn's work, *The Codebreakers: The Story of Secret Writing* (New York: Macmillan, 1967).

Zero-Knowledge Sudoku examples are taken from an August 3, 2006, post on my blog, *Computational Complexity* (http://blog.computationalcomplexity.org/2006/08/zero-knowledge-sudoku.html).

Works Cited

Whitfield Diffie and Martin Hellman, "New Directions in Cryptography," *IEEE Transactions on Information Theory* 22, no. 6 (November 1976): 644–54.

Craig Gentry, "Fully Homomorphic Encryption Using Ideal Lattices," in *Proceedings of the 41st Annual ACM Symposium on Theory of Computing* (New York: ACM, 2009), 169–78.

Ronald Rivest, Adi Shamir, and Leonard Adleman, "A Method for Obtaining Digital Signatures and Public-Key Cryptosystems," *Communications of the ACM* 21, no. 2 (February 1978): 120–26.

Chapter 9

My account of Richard Feynman's role in quantum computing draws on David Deutsch's work, "Quantum Computation," *Physics World*, January 6, 1992.

Works Cited

Charles Bennett and Gilles Brassard, "Quantum Cryptography: Public Key Distribution and Coin Tossing," *Proceedings of the IEEE International Conference on Computers, Systems, and Signal Processing* (Amsterdam: Elsevier, 1984), 175–79.

Charles Bennett, Gilles Brassard, Claude Crépeau, Richard Jozsa, Asher Peres, and William K. Wootters, "Teleporting an Unknown Quantum State via Dual Classical and Einstein-Podolsky-Rosen Channels," *Physical Review Letters* 70 (1993): 1895–99.

Lov Grover, "A Fast Quantum Mechanical Algorithm for Database Search," in *Proceedings of the Twenty-Eighth Annual ACM Symposium on Theory of Computing* (New York: ACM, 1996), 212–19.

Stephen Pincock, *Codebreaker: The History of Codes and Ciphers* (New York: Walker and Co., 2006), 151, for the Bennett and Brassard beach story.

Peter Shor, "Polynomial-Time Algorithms for Prime Factorization and Discrete Logarithms on a Quantum Computer," *SIAM Journal on Computing* 26 (1997): 1484–1509.

Chapter 10

Moore's law comes from Gordon Moore, "Cramming More Components onto Integrated Circuits," *Electronics* 38, no. 8 (April 19, 1965).

Watson's hardware is discussed in the IBM blog post, "What Runs IBM Watson and Why," by David Davidian.

The story of the shipping container comes from Marc Levinson, *The Box: How the Shipping Container Made the World Smaller and the World Economy Bigger* (Princeton, NJ: Princeton University Press, 2008).

Sources for Big Data Statistics

http://www.youtube.com/t/press_statistics

http://techcrunch.com/2011/03/14/new-twitter-stats-140m-tweets-sent-per-day-460k-accounts-created-per-day/

http://www.facebook.com/press/info.php?statistics

http://email.about.com/od/emailtrivia/f/emails_per_day.htm

http://public.web.cern.ch/public/en/lhc/Computing-en.html

http://space.about.com/od/telescopesandoptics/p/hubbleinfo.htm

http://webbtelescope.org/webb_telescope/technology_at_the_extremes/quick_facts.php

http://royal.pingdom.com/2011/01/12/internet-2010-in-numbers/

– – – *Index* – – –

Page numbers in *italics* indicate figures. Page numbers with n indicate notes.